Statistical Methods for Mediation, Confounding and Moderation Analysis Using R and SAS

Statistical Methods for Mediation, Confounding and Moderation Analysis Using R and SAS

Qingzhao Yu
Bin Li

CRC Press
Taylor & Francis Group
Boca Raton London New York

CRC Press is an imprint of the
Taylor & Francis Group, an **informa** business

A CHAPMAN & HALL BOOK

First edition published 2022
by CRC Press
6000 Broken Sound Parkway NW, Suite 300, Boca Raton, FL 33487-2742

and by CRC Press
2 Park Square, Milton Park, Abingdon, Oxon, OX14 4RN

CRC Press is an imprint of Taylor & Francis Group, LLC

© 2022 Taylor & Francis Group, LLC

ISBN: 978-0-367-36547-9 (hbk)
ISBN: 978-1-032-22008-6 (pbk)
ISBN: 978-0-429-34694-1 (ebk)

DOI: 10.1201/9780429346941

Publisher's note: This book has been prepared from camera-ready copy provided by the authors.

To the memory of Riming He.
To Zhiqiang Yu, Hailing Yu, Zhonghua Li,
Lingling Xia, Ari Li and Audriana Li.

Contents

Preface

Third-variable effect (TVE) refers to the intervening effect of a third-variable on the observed relationship between an exposure and an outcome. The third-variable effect analysis (TVEA) differentiates the effect from multiple third variables that explain the established exposure-outcome relationship. Depending on whether there is a causal relationship from the exposure to the third variable to the outcome, the third-variable effect can be categorized into two major groups: mediation effect where the causal relationship is assumed, and confounding effect where causal relationship is not presumed. The third variable is called a mediator or confounder respectively. A causal relationship can be established through randomized experiments, while statistical inference can be used to test the hypothesized associations among variables. In terms of the statistical inference on the third-variable effect, the mediation analysis and confounding analysis are equivalent, we call the analysis method the third-variable effect analysis in the book. Furthermore, the moderation effect is another important part of third-variable effect analysis. When the effects of a third-variable vary at different levels of another variable (called moderator), we use moderation analysis to differentiate the third-variable effect at each level of the moderator.

TVEA has been broadly applied in many fields. An example is to explore racial/ethnic disparities in health outcomes. Health disparities have been discovered in many areas, for example, the racial disparity in breast cancer survival. We found that among all US breast cancer patients, African Americans (AA) had an average lower survival rate than Caucasians (CA). We are interested to know 1) what factors (third-variables) could explain the racial differences, 2) if a risk factor is modifiable, how much could the racial disparity be improved by adjusting the risk factors so that they are equivalently distributed between the AA and CA and 3) Did the racial disparity change over time? If it did, which risk factor contributed to the change and how much change was brought by each essential risk factor. TVEA can be used to answer these questions in explaining disparities. In addition, when an intervention is implemented aiming at reducing certain health disparities, TVEA can be used to test whether the intervention is effective, and if it is effective, how it reduces the disparities. TVEA has been widely used in the areas such as social science, prevention study, behavior research, psychological analysis and epidemiology.

TVEA remains a challenge for researchers to consider complicated associations among variables and to differentiate individual effects from multiple third-variables. In this book, we introduce general definitions of third-variable effects that are adaptable to all different types of response (categorical or continuous), exposure, or third-variables. Using the method, multiple third-variables of different types can be considered simultaneously, and the indirect effect carried by individual third-variable can be separated from the total effect. Moreover, the derived third-variable analysis can be performed with general predictive models and multilevel models. That is, the relationship among variables can be modeled using not only generalized linear models but also nonparametric models such as the Multiple Additive Regression Trees and survival models. In addition, the TVEA can be performed with the hierarchical data structure in multilevel models. Furthermore, we use machine learning techniques in the TVEA to help identify important variables and estimate their effects simultaneously. We introduce the R package "mma" , "mmabig" and "mlma" for third-variable analysis and the SAS macros to call these packages from R.

The method proposed in the book can be easily adapted for practice if the reader has some basic knowledge on using R or SAS. Some linear regression, categorical data analysis, and/or survival analysis knowledge would be a great help to understand all contents in the book. In general, the book introduces methods and applications that can be adopted by anyone who has learned the introduction course of statistics or biostatistics. All R and SAS codes, and some supplementary materials are provided at the book website: http://statweb.lsu.edu/faculty/li/book/.

Last but the most, we would like to thank our family members, friends, colleagues and students for helping us make the book possible. We appreciate our advisers and professors at the Department of Statistics of The Ohio State University and our colleagues at the School of Public Health, Louisiana State University Health – New Orleans and the Department of Experimental Statistics at the Louisiana State University for their encouragements and advices. Our special thanks go to coauthors and students that have contributed to the book projects: Wentao Cao, Ying Fan, Paige Fisher, Ruijuan Gao, Joseph Hagan, Meichin Hsieh, Yaling Li, Kaelen L. Medeiros, Donald Mercante, Mary E. Moore, Richard Scribner, Xiaocheng Wu, Shuang Yang, Lu Zhang, Xiequn Zhang, Lin Zhu and many others. Finally, we would like to thank John Kimmel and David Grubbs, executive editors of Taylor & Francis, for the opportunity to work on this book. We are deeply grateful to Manisha Singh for the editorial assistance.

Symbols

Symbol Description

X The predictor(s) or exposure variable(s).

Y The response variable.

M The third-variables (confounder, mediator or moderator). The $X - Y$ relationship is established and we are interested in exploring the effects from intervention variables(M) in explaining the $X - Y$ relationship.

M_{-j} The third-variables (confounder, mediator or moderator) excluding the jth third-variable.

Z The covariates: Variables to explain Y or M other than X and M.

ATE The average total effect between an exposure and a response variable through all paths.

TE The total effect between an exposure and a response variable through all paths.

ADE The average direct effect between an exposure and a response variable after adjusting for other variables.

DE The direct effect between an exposure and a response variable after adjusting for other variables.

$DE_{\backslash M_j}$ The direct effect between an exposure and a response variable cutting the path through M_j.

AIE The average indirect effect through a third-variable between an exposure and a response variable.

IE The indirect effect through a third-variable between an exposure and a response variable.

RE The relative effect, which is defined as $\frac{DE}{TE}$ or $\frac{IE}{TE}$.

dom_x The domain of the random variable x.

u^* The infimum positive unit such that there is a biggest subset of dom_X, denoted as dom_{X^*}, in which any $x \in dom_{X^*}$ have that $x + u^* \in dom_X$, and that $dom_X = dom_{X^*} \cup x + u^* | x \in dom_{X^*}$.

$E_x g$ The expectation of $g(x)$.

$E_{x|y} g$ The conditional expectation of $g(x)$ given y.

$X \perp\!\!\!\perp Y | Z$ X and Y are independent given Z.

Π The conditional distribution of M given X.

1

Introduction

1.1 Types of Third-Variable Effects

A simple third-variable effect refers to the effect of a third-variable in explaining the observed relationship between an exposure variable and an outcome. Denote the third-variable as M, the exposure variable as X and the outcome as Y. The third-variable effects presented in this book are in one of the three formats: confounding, mediation and interaction/moderation. The third-variable is called a confounder or a mediator if it meets the following two conditions:

1. The third-variable is a risk factor to the outcome.

2. The third-variable is associated with the exposure variable.

Depending on whether the third-variable is in the causal pathway between the exposure and the outcome, the third-variable is called a mediator or a confounder. Figures 1.1 and 1.2 show the diagrams of a simple confounding effect and mediation effect separately. A line with an arrow indicates a causal relationship between the two related variables where the variable at the arrow head is caused by the variable at the other end of the line. A line without an arrow indicates that the two variables are associated but no causal relationship is established. The third-variable is called a mediator if there is a causal relationship between the exposure and the third-variable. Otherwise, the third-variable is a confounder.

Another type of third-variable effect is called the moderation/interaction effect, where at different level of the variable, the associations among the exposure, other third-variables, and the outcome are different. Figure 1.3 shows the diagram of a simple direct moderation/interaction effect. In a linear regression model, this association can be presented as an interaction term of the moderator and the exposure variable in explaining the outcome.

When there are multiple third-variables, the associations among variables can become more complicated. For example, the interaction effect can be between the moderator and another third-variable on the outcome; or between the moderator and the exposure variable on another third-variable. That is, there are different types of moderation effects. We discuss the different moderation effects and their inferences in Chapter 9.

DOI: 10.1201/9780429346941-1

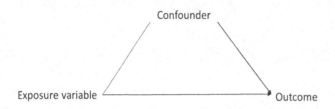

FIGURE 1.1
Confounding effect diagram.

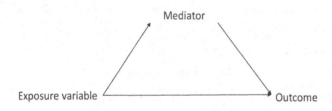

FIGURE 1.2
Mediation effect diagram.

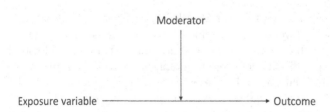

FIGURE 1.3
Moderation/interaction effect diagram.

1.2 Motivate Examples for Making Inferences on Third-Variable Effects

The major types of third-variable effect include confounding, mediation and moderation. In this section, we give a few examples showing the importance of making inferences on third-variable effects.

1.2.1 Evaluate Policies and Interventions

Often times, policies or interventions are designed to regularize human behavior or to treat certain diseases. For each intervention, there is usually a conceptual hypothesis of how it works. For example, an intervention for controlling weight may aim at helping people 1) increase physical activities, 2) have healthier food intake and 3) have better mood. To assess the effectiveness of the intervention, we have the following hypothesis to test:

1. The intervention is effective in reducing weight.

2. The intervention helps people reduce weight by increasing their physical activities.

3. The intervention helps people reduce weight by reducing their intake of calories.

4. The intervention helps people reduce weight by improving their mental health.

As measurements, the unit of reduced weight is the ounce of weight, which is the difference in the weight before and after the intervention. The amount of physical activities is measured by the average number of hours of physical activities per day. The mental status is measured by the Patient Reported Outcomes Measurement Information System (PROMIS) depression score, which is standardized to have a mean 50 and an overall standard deviation of 10. A higher depression score indicates a higher level of depression. All the measurements are taken before and after the one-month intervention. The calorie intake is measured by average daily calorie intake through a survey that records the food and drink taken in the past few days. There are other variables such as age, gender and education levels that are measured as control variables.

According to the study hypothesis, the conceptual model is shown in Figure 1.4. We are interested to estimate the effect from each different path connecting intervention and weight, and make inferences on whether each effect is significant or not. For the study hypothesis, we would like to see if the effect from each path is significantly different from 0. In addition, we want to compare and rank effects from different paths, so we know the strength and weakness of the intervention. Furthermore, we would like to know how each variable helps in changing the weight. For example, how efficient it is

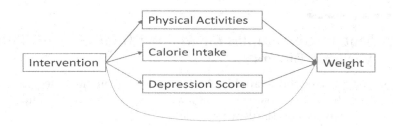

FIGURE 1.4
Conceptual model for weight intervention.

to increase physical activities in reducing weight. It is likely that the first half an hour's increase in physical activities might significantly help reduce weight, but the reduction may not be as substantial for the second half hour's increase of physical activity. That is, the relationship among variables might not be linear. There can be a bottleneck where further addition of physical activities cannot help much in reducing weights. We would like to know how much physical activity per day is most efficient and beneficial for individuals. All these questions can be answered by the third-variable effect analysis.

1.2.2   Explore Health Disparities

Health disparities are prevalent in the United States. Minorities suffer disproportionally from many poor health outcomes. For example, despite growing awareness of the negative health impacts of poor diet, physical inactivity and excess weight, the prevalence of obesity has increased dramatically in the United States (US). Hispanic and African Americans suffer disproportionately from obesity and related chronic diseases. Analysis of 2003-2006 National Health and Nutrition Examination Survey (NHANES) data shows that 44.3% of adult blacks and 30.2% of whites were obese. During the last three decades, research by the National Center for Health Statistics (NCHS) using NHANES shows that the US obesity rate doubled in adults and tripled in youth. Ethnic/racial disparities in obesity-related chronic diseases are also worsened. Diabetes prevalence increased 33.3% in whites, compared to 60.0% in blacks and 227.3% in Mexican Americans in the last decade. Mechanisms explaining these disparities are poorly understood. Both neighborhood and individual-level risk factors such as neighborhood walkability and individual physical activity behavior are shown to contribute to these racial/ethnic disparities.

By examining these potential explanatory factors from different levels (individual or neighborhood) jointly and differentiating the relative effects on racial/ethnic disparities, researchers are able to provide information to support targeted and precision medicine efforts. Moreover, such information is essential for public health officials and health care agencies in their efforts to develop efficient intervention strategies to reduce racial/ethnic disparities with limited resources. Third-variable effect analysis provides ways to identify important risk factors that explain the health disparities. It also helps differentiate and estimate effect of different factors from various levels (e.g. individual behavior and physical environment). With the analysis, we can also answer questions like: if we can manipulate such that the walkability for residential areas are equivalent for blacks and whites, what proportion of the racial difference in obesity could be reduced?

1.2.3   Exam the Trend of Disparities

An example of moderation effect is on monitoring the trend of disparity in health outcomes. Breast cancer is the most commonly diagnosed cancer for American women of all races. It is also the second leading cause of cancer death. Breast cancer has been categorized into subgroups for prognosis and treatment purposes. One common way of classifying breast cancer and recommending treatment is based on the expression of estrogen receptor (ER), progesterone receptor (PR), and human epidermal growth factor receptor 2 (HER 2) [7]. The subtype ER positive and/or PR positive (ER+/PR+) and HER2 negative (HER2-) breast cancer is the most common subtype, has the best prognosis and responds well to adjuvant endocrine therapy and/or chemotherapy. However, even within this subtype, patients have different recurrence risk and may respond to chemotherapy differently [54, 55].

Precision medicine has been developed significantly in today's cancer treatment. Oncotype DX® (ODX) is a genomic test that can differentiate ER+/PR+ and HER2- patients by the risk of recurrence to project prognosis and chemotherapy benefit. ODX test is based on 21-gene expression levels and it produces a recurrence score, a number between 0 and 100. A raised ODX score indicates a higher probability of cancer recurrence and more benefit from chemotherapy. The National Comprehensive Cancer Network (NCCN) cancer treatment guidelines published in 2008 recommended ODX test to patients with ER+/PR+, HER2-, and negative lymph node breast cancer to identify those that are more likely to benefit from chemotherapy. However, research shows that there are racial and ethnic disparities among breast cancer patients in terms of the survival rate, recurrence rate, and health-related quality of life [84, 86]. The disparity was also discovered in the use of ODX test [63, 40, 61]. Our previous work has shown that among all female breast cancer patients who were considered to be able to benefit from the ODX exam, non-Hispanic whites had a significantly higher rate of using the test, compared with non-Hispanic blacks [90]. In addition, the proportion of using ODX tests has been increasing over the last decade within both black and white patients. It is interesting to know whether the racial gap in ODX test has been reduced over time during the last decade. Furthermore, if there is a reduction in the gap, what factors contribute to this improvement? These questions can be answered through the third-variable-effect analysis.

1.3 Organization of the Book

For the rest of the book, we present the following contents. In Chapter 2, we review traditional methods of mediation and confounding analysis. We focus on two frameworks of research: regression models and the counterfactual framework. Then in Chapter 3, we review the advanced statistical modeling and machine learning techniques that are used in the third-variable analysis methods introduced in the book. Chapter 4 introduces general concepts of direct and indirect effects (Section 4.2), and proposes algorithms for third-variable effect analysis with generalized linear models, multiple additive regression trees, and smoothing splines. Readers who are mainly interested in the method application but not in the technique, can skip Sections 4.3 to 4.5. We introduce the R package *mma* for implementation of the proposed method in Chapter 5. The SAS macros to call the *mma* R package from SAS is also described. Section 5.3 illustrates the general method with multiple examples.

In Chapter 6, we discuss the assumptions requested for the method proposed in Chapter 4. We show the estimation bias and variances, and the sensitivity and specificity in identifying important third-variables if each of the assumptions is violated. In Chapter 7, we extend the general mediation and confounding analysis to multiple and multi-categorical exposure variables and to multivariate outcomes. Chapter 8 introduces the third-variable effect analysis for high-dimensional dataset, where we select important third-variables and estimate their effects simultaneously using regularized regression models. Readers who are not interested in the computing technique, can skip Section 8.2.

We further extend the general third-variable analysis method to account for interaction/moderate effect in Chapter 9. In the Chapter, we discuss types of different moderation effects and discuss the algorithm of making inferences on moderation effects. We analyze the trend in ODX diagnosis among breast cancer patients as an example for implementing the method.

In Chapter 10, we introduce the multilevel third-variable analysis using mixed additive regressions. We discuss definitions of third-variable effects for data coming with a hierarchical structure. In addition, we introduce the R package *mlma* that was compiled for the multilevel third-variable analysis. We show a real data example on using the method to explore racial disparity in body mass index accounting for environmental risk factors. Subsections 10.1.2 to 10.1.4 can be skipped if readers are not interested in the deduction of theorems.

Chapter 11 describes the use of Bayesian methods to implement the third-variable effect analysis. The advantages of using the Bayesian method are that we can use prior knowledge and information in addition to the data sets. Furthermore, the hierarchical data structure and spatial correlation among variables can be naturally implemented in the third-variable analysis through Bayesian modeling. We provide R and WinBugs codes and examples for the Bayesian third-variable analysis.

Last but not least, we discuss other issues in the third-variable analysis, which include the sample size and power analysis, how to explain the third-variable effects, and sequential third-variable analysis in Chapter 12.

Some supplementary materials are provided at the book website: http://statweb.lsu.edu/faculty/li/book/. In the site, we provide additional graphs from analysis that have been discussed in the book. Also, the website includes all R codes, SAS macros, and WinBUGS models that have been used in the book.

2

A Review of Third-Variable Effect Inferences

There are generally two frameworks for the third-variable analysis. One is under the general linear model setting, and the other is with the counterfactual framework. In this chapter, we review the two frameworks and conventional methods for third-variable analysis.

2.1 The General Linear Model Framework

For simplicity, we start with one third-variable and denote it as M. In addition, we denote the exposure variable as X and the outcome variable as Y. The purpose of the third-variable analysis is to estimate the indirect effect through the path $X \to M \to Y$ and the direct effect in the path $X \to Y$ as is shown in the conceptual model of Figure 2.1. Note that the arrowhead in the line from X to M can be removed to indicate that X and M are correlated. It is not necessary to assume X causes M.

2.1.1 Baron and Kenny Method to Identify a Third-Variable Effect

According to Baron & Kenny [4], three conditions are required to establish a third-variable effect:

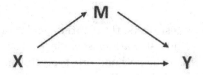

FIGURE 2.1
Simple third-variable conceptual relationship.

DOI: 10.1201/9780429346941-2

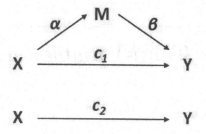

FIGURE 2.2
Simple third-variable relationship with coefficients.

1. the exposure variable (X) is significantly associated with the outcome variable (Y);

2. the presumed third-variable (M) is significantly related to X;

3. M significantly relates to Y controlling for X.

When both M and Y are continuous, and X is continuous or binary, assume the relationships among variables can be reasonably fitted by the following linear regression models (we ignore the intercepts for now):

$$M_i = \alpha X_i + \epsilon_{1i}; \tag{2.1}$$
$$Y_i = \beta M_i + c_1 X_i + \epsilon_{2i}; \tag{2.2}$$
$$Y_i = c_2 X_i + \epsilon_{3i}; \tag{2.3}$$

where ϵ_1, ϵ_2 and ϵ_3 are random error terms. In regression model 2.1, X is regressed on M. In equations 2.2 and 2.3, Y is regressed on Y with or without M. The coefficients are also shown in Figure 2.2.

The three conditions by Baron and Kenny [4] are transferred to the following three alternative hypotheses to be tested in the linear regression models:

1. $c_2 \neq 0$;

2. $\alpha \neq 0$;

3. $\beta \neq 0$.

Through these hypothesis tests, third-variable relationships and effects can be identified. However, no estimation is made on the third-variable effect. To make the inferences, two methods are typically used within the linear regression setting: the coefficient difference method (CD) and the coefficient product method (CP).

When the relationships are represented by linear regression models, the indirect effect is typically estimated by: CD – the difference in the coefficients of the exposure variable when it is regressed on Y with or without the third-variable M [38, 3, 41]; or CP – the product of the coefficient of X when regressed on M, and the coefficient of M in explaining Y controlling for X [49].

2.1.2 The Coefficient-Difference Method

Adopting coefficients from equations 2.1, 2.2 and 2.3, the CD method measures the total effect as c_2, the direct effect from X to Y as c_1, and the indirect effect from X to M to Y as $c_2 - c_1$. This is understood as that when the exposure X is used alone to estimate the outcome Y, all effects from X to Y are represented in the coefficient c_2. The assumption is that when one or more third-variables are added in the model, the effects of X on Y through the added third-variables are differentiated from c_2. Therefore, the difference between the coefficients of the exposure on the outcome with and without adjustment for the third-variable(s) is the effect from the exposure to the outcome through the third-variable(s). The uncertainties (variances and standard-deviation) in estimating the third-variable effects can be estimated through the Delta method or the bootstrap method.

If there are multiple third-variables, the effect from individual third-variable can be differentiated using the CD method. One way is to sequentially add in third-variables one-by-one. The difference in the coefficients for the exposure before and after adding each third-variable is the indirect effect for the third-variable. Benefits of the method include that the total effect (the coefficient of X when no third-variable is included in the linear model) and direct effect from exposure (the coefficient of X when all third-variables are included in the linear model) are invariant with the order of third-variables adding to the linear model, and the estimated total effect is always equivalent to the summation of the estimated direct effect and all estimated indirect effects. The disadvantage is that the estimates of indirect effects can change by the order of adding in the third-variables.

Another CD method can also be used to differentiate indirect effects from multiple third-variables. The indirect effect of a third-variable is estimated by the difference in the coefficients of the exposure variable in the linear model that includes only the exposure variable and the linear model that includes only the exposure variable and the third-variable. When estimating the indirect effect for one third-variable, all other third-variables are not included in the linear models. The direct effect is estimated as in the first method by the coefficient of the exposure variable when all third-variables are included in the linear model. One benefit of this version of CD is that the estimates are not influenced by the order of third-variables. In addition, the estimated total effect and the direct effect would not be influenced by the order of third-variables. However, the estimated total effect may not be the summation of all estimated indirect and direct effects. In addition, the method strictly depend on the assumption that given the exposure variable, effects of each third-variable on the outcome are independent of each other.

Overall, one benefit of using the CD method to estimate indirect effect is that when the third-variable has a nonlinear relationship with the response variable, the indirect effect is estimable. That is, the indirect effect is estimated through the difference of the coefficients of the exposure variable, $c_2 - c_1$, which does not depend on the format of the third-variable entering the linear model. Therefore, the third-variable can be transformed to enter the linear model. For the same reason, the CD method makes it possible to estimate the joint effect of a group of third-variables.

2.1.3   The Coefficient-Product Method

Using Equations 2.1 and 2.2, the direct effect from X to Y is c_1, the indirect effect through M is $\alpha \times \beta$, and the total effect is $c_1 + \alpha \times \beta$ by the CP method. When X changes by one unit, the third-variable M changes by α units. When M changes by one unit, Y changes by β units. Therefore, when X changes by one unit, Y changes by $\alpha \times \beta$ units through M. The indirect effect of M is interpreted as the amount of change in Y through M when X changes by one unit.

When the CP method is used for estimating third-variable effects, one of the major benefits is that the effect from multiple third-variables can be differentiated. For example, to find the indirect effect for the jth third-variable M_j, we first fit a linear model of the form 2.4 for the third-variable to get the coefficient of the exposure variable, $\hat{\alpha}_j$. Then the predictive model for the outcome is fitted that includes all third-variables and the exposure variable, Equation 2.5. From the complete predictive model, we obtain the estimated coefficient for M_j, denoted as $\hat{\beta}_j$. The product $\hat{\alpha}_j \times \hat{\beta}_j$ is the estimated indirect effect of M_j.

$$\hat{M}_{ij} = \hat{\alpha}_j X_i; \tag{2.4}$$

$$\hat{Y}_i = \hat{\beta}M_i + \hat{c}_2 X_i + \hat{\beta}_1 M_{i1} + \ldots + \hat{\beta}_j M_{ij} + \ldots; \tag{2.5}$$

Since the estimates of indirect effects involve the product of estimated coefficients from two linear models, the variances of the estimates of indirect effects cannot be directly estimated. The estimation of uncertainties can be obtained through the bootstrap method or through the Delta method. The estimated coefficients from each linear model are assumed to follow normal distributions. By the Taylor expansion, the variance of the estimated indirect effect is approximated as $\hat{\alpha}^2 \widehat{var(\beta)} + \hat{\beta}^2 \widehat{var(\alpha)}$, where $\hat{\alpha}$ and $\hat{\beta}$ are the estimated coefficients, and $\widehat{var(\beta)}$ and $\widehat{var(\alpha)}$ are the estimated variance for $\hat{\beta}$ and $\hat{\alpha}$, respectively.

Compared with the CD method, the CP method can easily separate indirect effect from multiple third-variables. But a limitation of the CP method is that the CP method cannot take into account of potential nonlinear relationships among variables easily. The coefficients are for the original format of the exposure variable and third-variables. If some variables are transformed to represent linear relationships, their coefficients can not be used directly in CD to estimate third-variable effects.

When all assumed equations are correct and there is only one third-variable in the model, it has been proved that $\alpha \times \beta = c_2 - c_1$, i.e. the inferences made by the CD method and the CP method are equivalent [49].

2.1.4   Categorical Third-Variables

When the third-variable is categorical, model 2.1 is not enough to describe the relationship between M and X. Usually a logistic regression is fitted for binary third-variables such that:

$$logit(P(M_i = 1)) \quad = \quad \alpha X_i. \tag{2.6}$$

Similarly, for a categorical variable M with $K + 1$ categories where $M = 0$ is used as the reference group, the third-variable can be binarized into K binary variables $M_i^T = (M_{i1}, \ldots, M_{iK})$ such that $M_{ik} = 1$ if $M_i = k$ and 0 otherwise. In the case, the relationship between M and X can be modeled by logistic regressions:

$$logit(P(M_{i1} = 1)) \quad = \quad \alpha_1 X_i;$$

$$\vdots$$

$$logit(P(M_{iK} = 1)) \quad = \quad \alpha_K X_i. \tag{2.7}$$

Note that we use the *logit* link here such that $logit(P(M_{ik} = 1)) = \log \frac{p(M_i = k)}{p(M_i = 0)}$. We can also use other links for categorical third-variables, for example, the probit link.

When M is categorical, the CD method can still be used to make inference on the indirect effect through M, where the estimated effect from M is still $c_2 - c_1$, where c_2 is the coefficient of the exposure in Equation 2.3 and c_1 is the coefficient of the exposure when M is included in the model:

$$Y_i = \sum_{k=1}^{K} \beta_k M_{ik} + c_1 X_i + \epsilon_{2i}.$$

As we discussed before, the CD method does not depend on the format of how the third-variable enters the model for the outcome, nor the association between X and M.

However, the indirect effect through M cannot be estimated by $\alpha \times \beta$ through the CP method for binary third-variables. Moreover, there is no straightforward function for estimating the indirect effect for multi-categorical third-variables by the CP method. The CP method does not work for the binary or categorical third-variable since with the model setting, when X changes by one unit, M does not change by an average α units as implied by $\alpha \times \beta$.

2.1.5 Generalized Linear Model for the Outcome

Now we look into the situation when the outcome variable is not continuous, and it is inappropriate to fit the outcome with linear regression models. Assume that the third-variable M is continuous, X is continuous or binary, and Equation 2.1 is still used to fit the relationship between X and M. We start with when Y is binary, and we use logistic regressions to fit the relationships between Y and other variables. In such case, linear regressions 2.2 and 2.3 are changed to the logistic regressions:

$$logit(P(Y_i = 1)) = \beta M_i + c_1 X_i; \qquad (2.8)$$

$$logit(P(Y_i = 1)) = c_2 X_i; \qquad (2.9)$$

In this situation, the CP method can still be used to estimate the indirect effect of M. The indirect effect is $\alpha \times \beta$. But the interpretation of the indirect effect is different. Instead of explaining the effect as the changing amount of the outcome when X changes by one unit, the indirect effect measures the change in the odds ratio of Y. The indirect effect from M can be interpreted as that when X increases by one unit, through the change in M, the odds of $Y = 1$ becomes $exp(\alpha \times \beta)$ times that of the original X.

On the other hand, the CD method cannot be used directly for the situation. In the logistic regression, there is not a random error term that includes all variances in the outcome (or transformed outcome) that cannot be explained by predictors. For linear regression, when different sets of predictors are used to fit the outcome, the variances of the outcome that cannot be explained by the predictors are included in the random error term. In the logistic regression, there is not a random error term. Therefore all variances in the outcome, $logit(P(Y = 1))$, are distributed among the estimated coefficients. When different sets of predictors are used to fit the logistic regression, the variances are distributed differently to the different sets of coefficients. Hence when using two sets of predictors to fit the outcome, even if a predictor is included in both sets, the variances distributed to the estimated coefficients of the same predictor are different. Thus the estimated coefficients of the predictor are not directly comparable.

In the logistic regressions 2.8 and 2.9, c_1 and c_2 have different scales. We cannot directly use $c_2 - c_1$ to estimate the indirect effect from M. A common practice is to standardize the scales for all coefficients and then estimate the third-variable effects based on the standardized coefficients. This practice becomes more complicated when multiple third-variables are involved. In addition, the interpretation of third-variable effects becomes more perplexing since the estimation is based on standardized coefficients but not on the original ones.

When using generalized linear models to deal with different types of outcomes and the CD method to estimate third-variable effects, it is important to understand the random terms in the generalized linear model to decide if the scales of estimated coefficients change with different sets of predictors.

2.1.6 Cox Proportional Hazard Model for Time-to-Event Outcome

When the outcome variable is time-to-event, denote the outcome as $Y = (T, \delta)$, where T is the shorter of the event time or censoring time, and δ indicates whether T is the event time ($\delta = 1$) or the censoring time ($\delta = 0$). A typical fitted model for the time-to-event outcome is the Cox-proportional hazard model. In the third-variable analysis, linear regressions 2.2 and 2.3 are changed to the following Cox models:

$$h(t) = h_{01}(t) \times exp(\beta M_i + c_1 X_i); \qquad (2.10)$$
$$h(t) = h_{02}(t) \times exp(c_2 X_i); \qquad (2.11)$$

The CP method can still be used to estimate the indirect effect of M where the indirect effect is $\alpha \times \beta$. Again, one should pay attention to the interpretation of the indirect effect. Instead of explaining the effect as the changing amount of the outcome when X changes by one unit, the indirect effect measures the multiplicative change in the hazard rate. The indirect effect from M can be interpreted as that when X increases by one unit, through the change in M, the hazard rate becomes $exp(\alpha \times \beta)$ times the original rate.

On the other hand, directly use the CD method to estimate the indirect effect of M is not correct. The important assumption for the cox model is the proportional hazard assumption, which says that all individuals have the same basic hazard function with a unique scale term that is decided by risk factors. The basic hazard function is the $h_0(t)$ function in the Cox model. When different sets of predictors are used to fit the Cox model (models 2.10 and 2.11), the basic hazard function can change – $h_{01}(t)$ and $h_{02}(t)$ are not necessarily the same. Therefore the c_1 in 2.10 and the c_2 in 2.11 cannot be directly compared. Moreover, it is not reasonable to assume both basic hazard functions are true.

Overall, although the generalized linear model and Cox model can deal with outcomes of different formats, neither CD nor CP method can both handle third-variables or outcomes of different formats, and differentiate effects from multiple third-variables [43]. The major problem for CD is that two models with different sets of predictors on the outcome need to be fitted for each third-variable. The problem with the CP method is that the interpretation of coefficients in the linear component of the generalized linear model depends on the random term of the model.

2.2 The Counterfactual Framework

Counterfactual framework is the other popular setting to implement the third-variable analysis [64, 57, 72, 1]. The concept of the potential outcome framework was first introduced by Donald B. Rubin in 1974 [65].

2.2.1   Binary Exposures

We first discuss the situation when the exposure variable is binary, taking the values of 0 or 1. Let X_i be the exposure for subject i. For example, $X_i = 0$ when subject i is in the control group and 1 for the treatment group. $Y_i(X)$ denotes the potential outcome if subject i is exposed to X, e.g. the post-treatment outcome. Based on these notations, $Y_i(1) - Y_i(0)$ is defined as the causal effect of the exposure on the response variable for subject i. To compare the change in outcome when the exposure changes from 0 to 1, only one of the responses, $Y_i(0)$ or $Y_i(1)$, is observed. It is impossible to estimate the individual causal effect since the estimation depends on a non-observable response. Holland (1986) proposed, instead of estimating causal effect on a specific subject, to estimate the average causal effect over a pool of subjects — $E(Y_i(1) - Y_i(0))$ [36]. If all subjects are randomly assigned to the groups of exposure, the average causal effect equals the expected conditional causal effect, $E(Y_i|X = 1) - E(Y_i|X = 0)$.

Denote $M_i(X)$ as the random third-variable M conditional on that subject i is exposed to treatment X. Let $m_i(1)$ or $m_i(0)$ be the observed value of M if subject i is actually assigned to the treatment or control group, respectively. Let $Y_i(x, m)$ be the potential outcome of subject i given that $X = x$ and $M = m$. It has been established that the total effect of X on Y when X changes from 0 to 1 is $Y_i(1, M_i(1)) - Y_i(0, M_i(0))$. The purpose of third-variable analysis is to decompose the total effect into direct effect and indirect effect(s). Namely, direct effect is the effect of X directly on Y, while indirect effect is the effect of X on Y through the third-variable M.

Robins and Greenland (1992) introduced the concepts of *controlled direct effect* and of *natural direct effect* [64]. The controlled direct effect is defined as $Y_i(1, m) - Y_i(0, m)$, where m is a realized value of M. The natural direct effect is defined as

$$\zeta_i(0) \equiv Y_i(1, M_i(0)) - Y_i(0, M_i(0)),$$

where $M_i(0)$ is the potential value of M when the exposure is fixed at $X = 0$. The difference between the controlled and natural direct effects is that the controlled direct effect is measured when M is fixed at m, whereas for the natural direct effect, M is random as if the actual exposure were 0. The difference between the total effect and natural direct effect is defined as the natural indirect effect of M, i.e.

$$\delta_i(1) \equiv Y_i(1, M_i(1)) - Y_i(1, M_i(0)).$$

That is, when X_i is fixed at 1, the change in Y_i when M_i changes from $M_i(1)$ to $M_i(0)$ is defined as the natural indirect effect. In comparison, the difference between the total effect and controlled direct effect cannot be generally interpreted as an indirect effect [76, 78].

The natural direct effect can be alternatively defined as $\zeta_i(1) \equiv Y_i(1, M_i(1)) - Y_i(0, M_i(1))$, where M_i is random given $X = 1$ but not at $X = 0$. The corresponding indirect effect is then defined as $\delta_i(0) \equiv Y_i(0, M_i(1)) - Y_i(0, M_i(0))$, where the exposure variable is fixed at $X = 0$. To ensure the uniqueness of the natural third-variable effect, it is necessary to assume that $\zeta_i(1) \equiv \zeta_i(0)$.

2.2.2 Continuous Exposure

Let $Y_i(X)$ denote the post-treatment potential outcome if subject i is exposed to X a continuous variable. To compare the change in outcome when the exposure changes from x to x^*, it is necessary to choose the two levels of comparison: $X = x$ and $X = x^*$ first. Again only one of the responses, $Y_i(x)$ or $Y_i(x^*)$, is observed. The causal effect of treatment on the response variable for subject i is defined as $Y_i(x) - Y_i(x^*)$. Instead of estimating causal effect on a specific subject, we can estimate the average causal effect over a pool of subjects — $E(Y_i(x) - Y_i(x^*))$. If the subjects are randomly assigned to $X = x$ or $X = x^*$, the average causal effect equals the expected conditional causal effect, $E(Y_i|X = x) - E(Y_i|X = x^*)$.

Denote $M_i(X)$ as the potential value of M when subject i is exposed to X. The total effect of X on Y when X changes from x to x^* is $Y_i(x, M_i(x)) - Y_i(x^*, M_i(x^*))$. The controlled direct effect and natural direct effects are similarly defined as for the binary exposure. The controlled direct effect is defined as $Y_i(x, m) - Y_i(x^*, m)$ and the natural direct effect is $\zeta_i(x) \equiv Y_i(x, M_i(x^*)) - Y_i(x^*, M_i(x^*))$. The natural indirect effect is therefore $\delta_i(x) \equiv Y_i(x, M_i(x)) - Y_i(x, M_i(x^*))$. The controlled indirect effect is not defined.

A common restriction for definitions of the controlled and natural direct effects is that the exposure levels x and x^* have to be preset. When the relationship among variables cannot be assumed as linear, the direct effect could be different at different x and x^* even if $x^* - x$ is fixed. Therefore, it is difficult to choose representative exposure levels.

2.2.3 Discussion

Using the counterfactual framework to explore third-variable effects has the benefit that the results are easy to be explained. However, the controlled indirect effect cannot be defined based on the definition of the controlled direct effect. For the natural direct effect, one has to assume that $\zeta_i(x) = \zeta_i(x^*)$ to ensure the uniqueness of the natural third-variable effects.

In addition, when the exposure variable is continuous and when the relationships among variables are not linear, it is important to choose the end points x and x^*, which can change the third-variable effects and their interpretations.

Perhaps the biggest problem of using the natural direct and indirect effects is the differentiation of multiple third-variables. When X is binary, there are two sets of natural third-variable effects. If one more third-variable is added, there could be four different sets of natural third-variable effects. Further, the estimation of the natural indirect effects depend on the order of estimation of the two third-variables. The number of definitions of natural third-variable effects exponentially increases with the number of third-variables when X is binary. The number is even bigger if the exposure is continuous or multi-categorical.

A more detailed review of mediation effect (third-variable) analysis methods can be found at [75]. The purpose of this book is to introduce a new method for third-variable analysis and software that is generally useful for multiple variables of different formats and can use parametric or nonparametric predictive models to fit relationships among variables. In Chapter 3, we review some statistical modeling and inference techniques that are used in the third-variable analysis methods that we present in the book starting in Chapter 4.

3

Advanced Statistical Modeling and Machine Learning Methods Used in the Book

In this chapter, we review some modeling and inference techniques that are used in the general third-variable analysis methods described in the book. We provide sample R codes for the topics covered in this section.

3.1 Bootstrapping

Bootstrapping is a useful resampling technique to estimate the uncertainty of statistic (e.g. standard error of a statistic) from sample data. Ideally, we can estimate the uncertainty of a statistic through repeated sampling from the population. The distribution of such sampled statistic is called the *sampling distribution*. However, in practice, we usually cannot repeatedly sample data from the population. The only information we have about the population is from the collected data. Therefore, bootstrapping uses random sampling *with replacement* to estimate the uncertainty of a statistic.

3.1.1 An Illustration of Bootstrapping

First, let's see an example of using bootstrapping to estimate the standard error of inter-quartile range (IQR), which is the difference between the third and first quartile. The data contain 200 independent observations generated from the standard exponential distribution (i.e. the probability density function is $f(x) = e^{-x}$). The R code below shows the repeated data sampling procedure from the population distribution. From each sampling, the IQR value is computed and the standard error is about 0.11. The left panel of Figure 3.1 shows the histogram of IQR from 5,000 samples collected from the population. The density plot (grey curve) can be viewed as an approximation of the sampling distribution of IQR.

```
> iter = 5000 #Repeat sampling 5000 times
> iqr.pop = rep(0,iter)
```

DOI: 10.1201/9780429346941-3

```
> for (i in 1:iter){
+   set.seed(i+1)
+   iqr.pop[i] = IQR(rexp(200,rate=1))
+ }
> sd(iqr.pop)
[1] 0.1137913
```

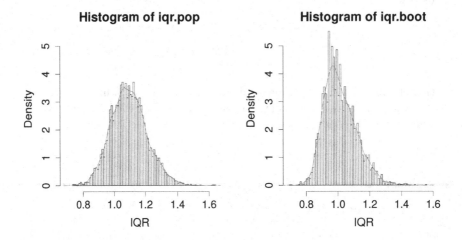

FIGURE 3.1
The sampling and bootstrap distributions of IQR.

The right panel of Figure 3.1 shows the histogram of IQR based on 5,000 bootstrap samples from a particular random sample. The estimated standard error of the IQR from the bootstrap samples (i.e. about 0.10) is close to the one estimated by directly sampling from the population distribution. The density plot, the right panel of Figure 3.1, has similar shape and range to that on the left panel. Bootstrapping is particularly useful when the analytical form of the sampling distribution is not available.

```
> iqr.boot = rep(0,iter)
> for (i in 1:iter){
+   set.seed(i+1)
+   iqr.boot[i] = IQR(sample(x,size=200,replace=T))
+ }
> sd(iqr.boot)
[1] 0.1013573
```

3.1.2   Bootstrapping for Linear Regression

Bootstrapping can be used in more complicated situations. Let's see an example of using bootstrapping in linear regression problem. Duncan's occupational prestige data contains the prestige and other characteristics of 45 US occupations in 1950. The dataset is included in the *car* package in R. The response variable, `prestige` is the percentage of the respondents in a survey who rated the occupation as "good" or better in prestige. The simple linear regression model below explores the linear association between two variables: prestige and income. The left panel of Figure 3.2 shows the scatter plot of the data with the fitted line from the regression model. The right panel presents the residual plots. We see there are some outlying residuals on the top right corner. From the model outputs, we see that although the slope is highly significant, the residual is not normally distributed from the Shapiro-Wilk's test results. This makes the test results on the slope questionable. Alternatively, we can use bootstrapping to make statistical inference on the model statistic, for example, to get the confidence interval (CI) of the slope.

```
> library(car)
> attach(Duncan)
> lr.fit <- lm(prestige ~ income)
> shapiro.test(lr.fit$resid)

        Shapiro-Wilk normality test

data:  lr.fit$resid
W = 0.93505, p-value = 0.0142
> summary(lr.fit)
Coefficients:
            Estimate Std. Error t value Pr(>|t|)
(Intercept)   2.4566     5.1901   0.473    0.638
income        1.0804     0.1074  10.062 7.14e-13 ***
---
Signif. codes:  0 '***' 0.001 '**' 0.01 '*' 0.05 '.' 0.1 ' ' 1
Residual standard error: 17.4 on 43 degrees of freedom
Multiple R-squared:  0.7019,     Adjusted R-squared:  0.695
F-statistic: 101.3 on 1 and 43 DF,  p-value: 7.144e-13
```

The following R code generates the 95% bootstrap interval for the slope $\hat{\beta}_{income}$ using *bootstrapping pairs* approach. Namely, we estimate the slope $\hat{\beta}_{income}$ based on each bootstrap sample by randomly sampling the 45 occupations with replacement. The bootstrap samples are drawn many times (e.g. 5000 times here). The lower and upper 2.5 percentile from the bootstrap estimates are the lower and upper bounds for the bootstrap interval of the slope $\hat{\beta}_{income}$. From the R output, we see that the bootstrap interval is different from the t-test based CI, especially in the upper limit. This is probably due to the existence of an outlying residual appeared at the upper right corner of the residual plot. Figure 3.3 shows the empirical distribution of $\hat{\beta}_{income}$ from bootstrapping. The estimated 95% bootstrap interval is shown as the

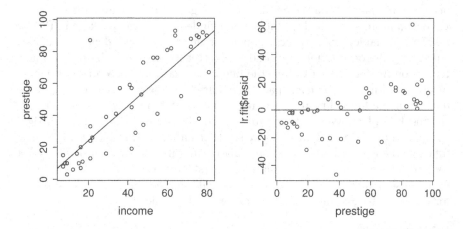

FIGURE 3.2
Left panel: the scatter plot and fitted line for Duncan's prestige data. Right
panel: the residual plot.

two red vertical bars at the bottom. The blue vertical line in the middle is the
estimated slope using the original dataset.

```
> beta.boot <- rep(0,iter)
> for (i in 1:iter){
+    set.seed(i)
+    ind <- sample(1:45,size=45,replace=T)
+    fit <- lm(prestige~income,data=Duncan[ind,])
+    beta.boot[i] <- fit$coef[2]
+ }
> #95% bootstrap interval
> quantile(beta.boot,prob=c(0.025,0.975))
     2.5%      97.5%
0.8611574 1.2699147
> #95% t-test based CI
> confint(lr.fit,"income")
          2.5 %    97.5 %
income 0.8638599 1.296919$
```

3.2   Elastic Net

Recently, several shrinkage and regularization methods, such as LASSO [73]
and Elastic Net [94], have been proposed as alternatives to the classic ordinary

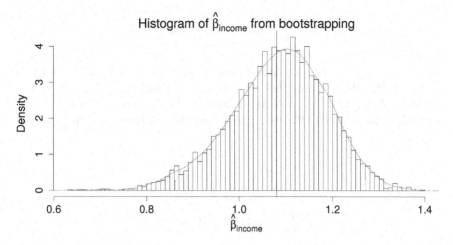

FIGURE 3.3
The bootstrap distributions of the slope $\hat{\beta}_{income}$.

least squares (OLS) estimates. Consider a linear regression model: given p predictors x_{ij}, $j = 1, 2, \ldots, p$ and the response y_i for the ith observation, $i = 1, 2, \ldots, n$. Suppose the response y is a linear function of p predictors x_j with an additive random error ϵ:

$$y_i = \beta_0 + \sum_{j=1}^{p} \beta_j x_{ij} + \epsilon_i. \tag{3.1}$$

The OLS estimator $\hat{\beta}^{ols}$ minimizes

$$SSE = \sum_{i=1}^{n} \left(y_i - \hat{\beta}_0 - \sum_{j=1}^{p} x_{ij} \hat{\beta}_j \right)^2. \tag{3.2}$$

Based on the Gauss-Markov Theorem, we know OLS estimator is the *best linear unbiased estimator* (BLUE) of the regression coefficients under the assumptions that the random errors $\{\epsilon_i\}_1^n$ are zero-mean, constant variance and uncorrelated. The word "best" in BLUE means the OLS estimator achieves the smallest mean squared errors (MSE)

$$MSE(\hat{\beta}) = \sum_{j=1}^{p} E(\hat{\beta}_j - \beta_j)^2 \tag{3.3}$$

$$= \sum_{j=1}^{p} E[\hat{\beta}_j - E(\hat{\beta}_j)]^2 + \sum_{j=1}^{p} E[E(\hat{\beta}_j) - \beta_j]^2$$

$$+ \sum_{j=1}^{p} 2E(\hat{\beta}_j - E(\hat{\beta}_j))(E(\hat{\beta}_j) - \beta_j) \tag{3.4}$$

$$= \sum_{j=1}^{p} Var(\hat{\beta}_j) + \sum_{j=1}^{p} [Bias(\hat{\beta}_j)]^2 \tag{3.5}$$

among all the unbiased estimators. However, that doesn't mean the OLS estimator is the "best" among all the estimators. From Eq. 3.5, we know that the MSE of $\hat{\beta}^{ols}$ is the sum of its variance, since the second term in Eq. 3.5 is zero for the unbiased OLS estimator. Note that all the cross-product terms in Eq. 3.4 are equal to zero

$$\sum_{j=1}^{p} E(\hat{\beta}_j - E(\hat{\beta}_j))(E(\hat{\beta}_j) - \beta_j)$$

$$= \sum_{j=1}^{p} (E(\hat{\beta}_j) - \beta_j) E(\hat{\beta}_j - E(\hat{\beta}_j))$$

$$= \sum_{j=1}^{p} (E(\hat{\beta}_j) - \beta_j) \times 0 = 0.$$

It is known that OLS estimators can be highly variable when some regressors are correlated to others, which may result in poor prediction performance. In addition, OLS estimators do not select variables. It is often the case that some of the input variables in the regression model are irrelevant to the response. By removing the irrelevant variables from the model, we can easily interpret the results. This is particularly crucial for the high-dimensional problems, where the number of predictors p is large.

3.2.1   Ridge Regression and LASSO

The main idea of using shrinkage estimators is to sacrifice a little bias in order to substantially reduce the variance so that the overall performance is improved (i.e. MSE is reduced). The shrinkage estimators use a technique called *regularization*, which was originally proposed by Andrey Tikhonov, a Russian mathematician, to solve the ill-posed problem, where the solutions are not unique.

Ridge regression [17] is one of the oldest shrinkage methods in the regression analysis. Ridge estimator $\hat{\beta}^R$ shrinks the regression coefficients towards zero by adding an L_2 penalty on the coefficients:

$$\hat{\beta}_\lambda^R = \arg\min_{\beta} SSE + \lambda \sum_{j=1}^{p} \beta_j^2, \tag{3.6}$$

where λ is a non-negative tuning parameter that controls the bias and variance in Eq. 3.4. For large λ, the ridge penalty dominates the objective function and shrinks the coefficients towards zero, which increases the squared bias and decreases the variance. It is known that ridge regression performs well when at least some of the regressors are highly correlated. In addition, the ridge estimator is unique even with $p > n$. However, ridge regression does not select variables. Tibshirani (1996) proposed *lasso*, another popular shrinkage estimator, that can select variables. Lasso estimator $\hat{\beta}^L$ shrinks the regression coefficients towards zero by adding an L_1 penalty on the coefficients:

$$\hat{\beta}_\lambda^L = \arg\min_\beta SSE + \lambda \sum_{j=1}^p |\beta_j|. \tag{3.7}$$

For large value of λ, some of the Lasso coefficients can be exactly zero. The Figure 3.4 below shows the solution paths for ridge and lasso in a simulation example, where the data are generated from the true model: $y = X\beta + \epsilon$ with $\beta = (3, 1.5, 0, 0, 2, 0, 0, 0)$, $\epsilon \sim N(0, 3^2)$, and $X \sim N(0, \Sigma)$. The variance-covariance matrix Σ was set to be $\mathrm{Cov}(i, j) = 0.5^{|i-j|}, \forall i \neq j$. This simulation example was originally used in [94]. For both ridge and Lasso, the optimal values for the tuning parameter λ can be chosen to minimize the cross-validation (CV) error. The vertical dash lines in Figure 3.4 show the estimated regression coefficients using the optimal values of $log(\lambda)$ that minimizes the 10-fold CV error. From Figure 3.4, we see the ridge estimator does not elect the variables, while lasso does. The estimated Lasso solution contains four non-zero coefficients. Note the numbers on the top axes are the numbers of non-zero coefficients in the solution. Studies have shown that ridge works well under the "dense" situation, where majority of the regression coefficients are non-zero, and Lasso works well under the "sparse" situation, where only a few variables have non-zero coefficients.

```
> library(MASS)
> S = matrix(0,8,8)
> for (i in 1:8){
+    for (j in 1:8){
+       S[i,j]=0.5^abs(i-j)
+    }
+ }
> n = 100
> set.seed(1)
> x = mvrnorm(n,mu=rep(0,8),S)
> b = c(3,1.5,0,0,2,0,0,0)
> y = x%*%b+rnorm(n,sd=3)
> library(glmnet)
> ridge.fit = glmnet(x,y,alpha=0)
> lasso.fit = glmnet(x,y,alpha=1)
> set.seed(1)
> ridge.cv = cv.glmnet(x,y,alpha=0)
> lasso.cv = cv.glmnet(x,y,alpha=1)
```

```
> ridge.cv$lambda.min #optimal lambda minimizes CV error
[1] 0.5530578
> lasso.cv$lambda.min
[1] 0.3092041
> par(mfrow=c(1,2))
> plot(ridge.fit,xvar="lambda",main="Ridge")
> abline(v=log(ridge.cv$lambda.min),lty=2)
> plot(lasso.fit,xvar="lambda",main="Lasso")
> abline(v=log(lasso.cv$lambda.min),lty=2)
```

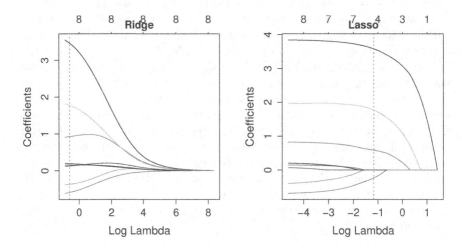

FIGURE 3.4
The solution paths for ridge and Lasso.

3.2.2   Elastic Net

Although Lasso can select variables, it is known that when $p > n$, Lasso can only select at most n variables. In addition, when some variables are grouped among which the pairwise correlations are high, Lasso tends to select only one variable to represent the group and ignore the rest. In order to fix these problems, [94] proposed elastic net, which combines the L_1 and L_2 penalties from lasso and ridge, respectively:

$$\hat{\beta}_{\lambda,\alpha}^{Enet} = \arg\min_{\beta} SSE + \lambda \left[\alpha \sum_{j=1}^{p} |\beta_j| + \frac{1}{2}(1-\alpha) \sum_{j=1}^{p} \beta_j^2 \right]. \qquad (3.8)$$

When $\alpha = 1$, elastic net becomes lasso. When $\alpha = 0$, elastic net is ridge. Studies have shown that elastic net often outperforms lasso and ridge, and can select groups of correlated variables. Figure 3.5 below the solution path and the 10-fold CV error plot for the elastic net model with $\alpha = 0.9$. The CV error plot shows the CV error together with the standard error bars. There are two vertical dash lines in the CV plot. The left one shows the optimal value of λ which gives the smallest CV error. The right one shows the optimal value of λ based on the 1-SE criterion. Namely, the largest value of λ such that the CV error is within one standard error of the minimum. The former estimate is mainly for the prediction purpose, while the 1-SE criterion is usually used for interpretation purposes.

```
> enet.fit = glmnet(x,y,alpha=0.9)
> set.seed(6)
> enet.cv = cv.glmnet(x,y,alpha=0.9)
> enet.cv$lambda.min
[1] 0.2852297
> plot(enet.fit,xvar="lambda",main="Elastic net")
> abline(v=log(enet.cv$lambda.min),col="red",lty=2)
> plot(enet.cv)
```

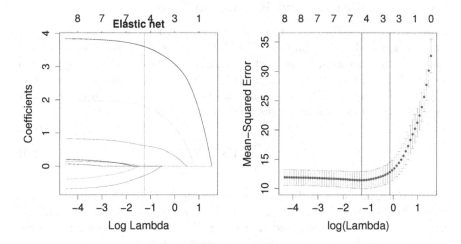

FIGURE 3.5
The solution path (left) and 10-fold CV error plot (right) for elastic net (with $\alpha = 0.9$).

3.3   Multiple Additive Regression Trees

Boosting is one of the most successful ensemble learning schemes introduced at the end of the 20th century. The idea of boosting was originally proposed by Schapire [69] with the purpose to produce an accurate prediction rule by combining a set of less accurate rules generated from *weak learners*. The AdaBoost (Adaptive Boosting) algorithm proposed by Freund and Schapire [21] is generally considered as the first step of putting the boosting idea into practice. Friedman et al. [24] provided a statistical interpretation of AdaBoost as a *forward stagewise* modeling procedure of fitting an additive logistic regression model using the exponential loss. Note that the *forward stagewise* procedure updates the model iteratively by adding new terms into the model without changing the previously added terms. The closely-named *forward stepwise* procedure adds new term as well as updates the previously fitted model at each iteration. While both stagewise and stepwise approaches are greedy algorithms to fit additive models, the latter overfits much faster than the former. Later, Friedman [22] presented a generic gradient tree-boosting algorithm, called "multiple additive regression trees" (MART), which allows optimization of any differentiable loss function through functional gradient descent. Due to its superior empirical performance over a wide range of problems, MART is considered as one of the best off-the-shelf procedures for data mining [31].

3.3.1   Classification and Regression Tree

Since MART algorithm uses tree as the base learner, we briefly review the classification and regression tree (CART [10]), one of the most popular machine learning methods, in this section. CART partitions the input space into non-overlapping rectangular regions called *leaf*. Tree fits a simple model (e.g. majority class for classification and constant for regression) on each leaf. A regression tree $T(\mathbf{x})$ can be represented as the following equation:

$$T(\mathbf{x}) = \sum_{s=1}^{S} \beta_s \mathbf{I}(\mathbf{x} \in R_s), \tag{3.9}$$

where $\{R_s\}_{s=1}^{S}$ is a set of disjoint terminal regions, $\mathbf{I}(\cdot)$ is the indicator function and S is the total number of leaves (also called tree size). The splitting variables, split points and fitted values (e.g. β_s) are chosen to minimize some pre-defined loss function, such as SSE for regression tree.

Figure 3.6 shows a fitted regression tree on Duncan's prestige data, which is described in Section 3.1.2. The size of the tree is three, since it has three leaves (node 3, 4 and 5). Node 1 is called the *root*. The number of edges from the root down to the furthest leaf is called the *maximum depth*, which is two in this case. Two splitting variables are used in the fitted regression tree: `income` and `type`, which contains three categories of occupations: `prof`: professional and managerial; `wc`: white-collar; `bc`: blue-collar. The rightmost leaf (node 5) contains 18 professional occupations, which have the highest fitted prestige value (i.e. about 80%). On the other hand, the leftmost leaf (node 3) contains 17 non-professional (blue- and white-collar) and low-income occupations, which have the lowest fitted prestige value (about 16%). The R code below uses `rpart` and `partykit` packages to fit and plot a regression tree in R.

```
> #Recursive Partitioning And Regression Trees
> library(rpart)
> library("partykit") #toolkit for visualizing trees
> tree.fit = rpart(prestige~income+type,data=Duncan)
> plot(as.party(tree.fit),type="simple")
```

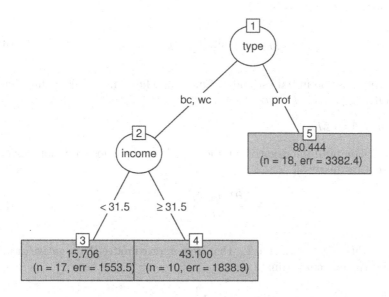

FIGURE 3.6
A fitted regression tree model for Duncan's prestige data.

Tree has several advantages over traditional statistical models. First, tree model does not require any distributional assumptions on the data. Second, tree can handle mixed type of data, both continuous and categorical inputs, as well as the missing values in the input variables. Third, tree model can fit non-additive behavior (i.e. complex interaction effects among the input variables) and nonlinear effects with relatively little input required from the investigator. Fourth, tree model can be visualized through the tree diagram. Lastly, tree can be easily interpreted as IF/THEN language. Since MART uses trees as base learners, it naturally inherit the important advantages from CART, especially appropriate for mining less than clean data. On the other hand, tree model often has high variance. A small change of data set can result in a big change of the tree structure. Another major problem of tree is its relatively poor prediction performance, which can be overcome by combining multiple trees in MART.

3.3.2   MART Algorithm

MART fits an additive model, in which individual terms are trees, in a sequential fashion that best reduces the loss function through functional gradient descent. Let $\{y_i, \mathbf{x}_i\}_1^n = \{y_i, x_{i1}, \ldots, x_{ip}\}_1^n$ be n observations and $L(y, f)$ be any differentiable loss function. MART first finds the constant estimate f_0 that minimizes the loss function

$$f_0 = \arg\min_f \sum_{i=1}^n L(y_i, f),\qquad(3.10)$$

and uses it as the initial estimated function. Then MART algorithm iterates the following steps M times.

For $m = 1$ to M:

Step 1 Compute the negative gradient $\{g_i\}_{i=1}^n$ of the loss function w.r.t. the current estimate f_{m-1}.

$$g_i = -\left[\frac{\partial L(y_i, f)}{\partial f}\right]_{f = f_{m-1}(\mathbf{x}_i)}.\qquad(3.11)$$

Step 2 Fit a regression tree T_m that best approximates the negative gradient under the current estimate

$$T_m(\mathbf{x}) = \arg\min_T \sum_{i=1}^n (T_m(\mathbf{x}_i) - g_i)^2\qquad(3.12)$$

giving a set of fitted terminal regions $\{R_s\}_{s=1}^{S_m}$.

Step 3 Find the fitted values for all the terminal regions (i.e. $\{R_s\}_{s=1}^{S_m}$) of the regression tree T_m to minimize the loss function $L()$.

$$\beta_s = \arg\min_\beta \sum_{x_i \in R_s} L(y_i, f_{m-1} + \beta)^2, \quad \forall s = 1, \ldots, S_m.\qquad(3.13)$$

Step 4 Update the current estimate by adding the fitted tree T_m into the ensemble

$$f_m(\mathbf{x}) = f_{m-1}(\mathbf{x}) + \sum_{s=1}^{S_m} \beta_s \mathbf{I}(x \in R_s).\qquad(3.14)$$

3.3.3 Improvement of MART

It has been shown that the prediction performance of MART can be substantially improved by two techniques, *regularization* and *subsampling*. The former is to downweight each tree by multiplying a small positive number ν, also called *shrinkage* parameter

$$f_m(\mathbf{x}) = f_{m-1}(\mathbf{x}) + \nu \sum_{s=1}^{S_m} \beta_s \mathbf{I}(x \in R_s). \tag{3.15}$$

Studies have shown that using small shrinkage value (e.g. $0 < \nu \ll 1$) prevents MART from overfitting and improves the prediction performance. The latter is to use a random subset of training samples to construct each tree, see [9] and [23]. In [9], Breiman proposed to use bootstrap samples to fit the base learner within each iteration. Friedman [23] proposed a slightly different approach, called *stochastic gradient boosting*, to incorporate randomness into the procedure. Specifically, at each iteration a subsample of the training data is drawn at random *without replacement* to fit the base model.

In practice, the optimal number of trees M can be estimated by monitoring the prediction performance through an independent validation set, cross-validation (CV) and out-of-bag (OOB) samples, which are the samples not used to construct a particular tree. Although using smaller shrinkage parameter ν usually gives better prediction performance, it gives rise to larger M (i.e. the number of trees).

Besides the number of trees M and the learning rate ν, MART algorithm has another tuning parameter: the size of the tree (i.e. the number of leaves of a tree). Studies have shown that using very large trees in the MART ensemble substantially degrades the prediction performance and increases computation. On the other hand, using simple trees (e.g. stump, tree with only a single split) may not complicate enough to catch the interaction between input variables. A convenient way is to restrict all the trees to the same maximum depth, which is directly related to the highest order of interaction that a tree can explain. For example, by setting the tree depth to 1, the MART model is additive with only main effects. With depth equals 2, two-variable interaction effects are allowed, and so on.

3.3.4 A Simulation Example

Let's see a MART application on a simulation example with ten X variables and numeric response Y. We simulated data from the following true model:

$$y = 0.5x_1 + 0.17x_2^2 + 140\phi(x_3, x_4) + \epsilon, \ \epsilon \sim N(0, \sigma^2) \tag{3.16}$$

where predictors are generated from a multivariate normal distribution with *mean* = 0 and *variance* = 3. All the pairwise correlations are fixed at 0.5. The function $\phi()$ is the probability density function for a bivariate normal distribution with *mean* $= (0, 0)$, *variance* $= (6, 4)$ and *covariance* $= 2$. Only the first four variables are relevant to the response Y. The coefficients for three terms (i.e. 0.5, 0.17 and 140) are chosen so that each term has approximately the same contribution (i.e. variance for each term is about the same). The value of σ^2 is chosen to have the signal-to-noise ratio is 10 (i.e. $var(y) = 11\sigma^2$).

An R implementation of MART is available in gbm package, which uses the stochastic gradient boosting algorithm on half of the subsample with shrinkage parameter $\nu = 0.1$ as the default setting. In addition, the gbm package allows using several loss functions in MART, such as absolute loss (L_1 loss) and Huber loss in regression. We run the MART in library(gbm) using the following code.

```
> library(gbm)
> set.seed(1)
> fit.mart=gbm(y~.,distribution="gaussian",data=data,
        n.tree=100,interaction.depth=3,shrinkage=0.01,
        train.fraction=0.8,bag.fraction=0.5,verbose=F)
> best.iter=gbm.perf(fit.mart, plot.it=FALSE,
                     oobag.curve=F, method="OOB")
> while(fit.mart$n.trees-best.iter<50){
+    fit.mart <- gbm.more(fit.mart, 100)
+    best.iter <- gbm.perf(fit.mart, plot.it=FALSE,
+    oobag.curve=F, method="OOB")
+ }
> best.iter #The optimal number of trees
[1] 952
> fit.mart$n.tree #Total number of trees
[1] 1100
```

Here are some remarks on the above code.

- The distribution option specifies the distribution of the response and the loss function used in MART. The squared error loss in regression corresponds to gaussian distribution. The absolute loss is laplace. The 0-1 outcomes in logistic regression is bernoulli, etc.

- The interaction.depth specifies the maximum depth for each tree, which is also the highest level of variable interactions allowed.

- The shrinkage parameter ν is specified in shrinkage term.

- train.fraction=0.8 indicates the first 80% observations are used to fit the gbm model and the remainder are test observations.

- We set the bag.fraction=0.5, which means half of the training set observations are randomly selected to build the next tree (i.e. subsampling) in the expansion. This introduces randomness into the model fitting. By setting the random seed in set.seed, we can ensure that the results can be exactly reproduced.

- The OOB approach is used to find the optimal number of iterations. Namely, 100 trees is fitted firstly by setting `n.tree=100`. If the optimal number of iterations `best.iter`, evaluated by the `gbm.perf` function, is within 50 to the total number of iterations (e.g. 100), 100 more trees are added to the ensemble. Otherwise, stop the process.

- If the cross-validation approach is preferred for choosing the optimal values of the parameters, use the `cv.folds` option (e.g. `cv.folds=5` performs a 5-fold CV.)

3.3.5 Interpretation Tools for MART

Although single trees are highly interpretable, it is hard to interpret the MART ensemble, which includes hundreds of trees, in the same way. A few interpretation tools have been developed to shed light on the underlying data generating mechanism.

```
> #Computes the relative influence of each variable
> summary(fit.mart,n.trees=best.iter)
> #Plot the performance measure in MART
> gbm.perf(fit.mart,plot.it=T,oobag.curve=F,method="OOB")
```

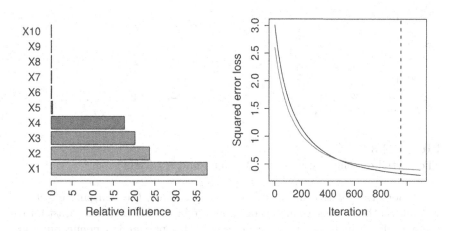

FIGURE 3.7
The relative variable influence (left) and performance plot (right) for MART.

Figure 3.7 shows the relative variable influence for all ten X variables in the above simulation example. We see that the first four variables dominate the influence on the response, while the rest of the six variables have the minimum effects on prediction, which agrees with the underlying data generating model described in Eq. 3.16. In practice, we can select variables based on the relative variable influence measures from MART. Note that the relative influence of each variable is scaled so that the sum of the relative influence is 100. A higher relative influence measure indicates stronger influence of the variable on the response. The right panel shows the performance measures on the training (black curve) and test/validation (grey curve) sets. The dash vertical line shows the optimal number of trees based on the OOB performance. It is known that the OOB approach generally underestimates the optimal number of trees in MART, although the predictive performance is reasonably competitive. Cross-validation usually provides a better estimate of the number of trees. However, CV is much more computationally intensive, especially for the large data sets.

```
> #Partial dependence plots for variable 1&2
> plot(fit.mart,i.var=1,n.trees=best.iter)
> plot(fit.mart,i.var=2,n.trees=best.iter)
```

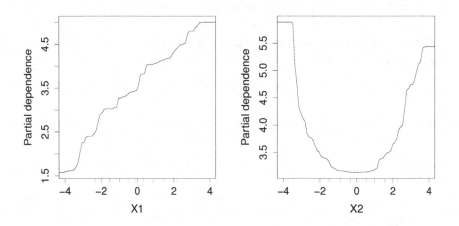

FIGURE 3.8
The partial dependence plots for X_1 (left) and X_2 (right).

Besides the relative influence for each predictor, the marginal effect of a variable on the response can also be calculated from MART that adjust for the average effects of all the other variables using the partial dependence function. Figure 3.8 shows the partial dependence plots for $X1$ (left) and $X2$ (right), the leading two most influential variables. The tick marks at the bottom shows the distribution in deciles (e.g. 10th, 20th, ... percentiles) of the corresponding predictor. We see the partial dependence plots reflect the underlying true functional effects for both variables. For example, the partial dependence plot for $X1$ is monotonically increasing, being nearly linear over the major part of data. Note that the data density is low near the edges, which causes the curves to be less determined at these regions.

Friedman and Popescu [27] proposed a randomization test to examine the interaction effects among predictors on the response variable. The `interact.gbm` function in `gbm` package is used to compute the H-statistics to assess the strength of variable interactions. The code below computes the strength of the two-way interaction effects between $X4$ and other 9 predictors. We see the interaction effect between $X3$ and $X4$ is much larger than that for others. Figure 3.9 shows the joint effects of $X3$ and $X4$ on the response. The peak is around the origin, which agrees with the underlying true function (i.e. ϕ is a bivariate normal PDF with zero mean). Note that, partial dependence function of a selected variable is useful only when the selected variable does not have any strong interaction with other variables in the data set.

```
> indx=1:10[-4]
> interaction.x4=rep(0,9)
> for (j in 1:9){
+    interaction.x4[j]=interact.gbm(fit.mart,data,
+           i.var=c(indx[j],4),n.trees=best.iter)
+ }
> round(interaction.x4,d=4)
[1] 0.0430 0.0512 0.2034 0.0008 0.0015
[6] 0.0009 0.0008 0.0014 0.0034
> plot(fit.mart,i.var=3:4,n.trees=best.iter)
```

3.4 Generalized Additive Model

Generalized additive model (GAM) is a popular and powerful statistical modeling technique which enjoys three key features: interpretability, flexibility and regularization. We will introduce GAM and its main component: smoothing splines in this section.

3.4.1 Generalized Additive Model

Generalized additive model [32] fits an additive model with flexible smooth basis functions on a univariate response. For regression problem, GAM has the form

$$E(Y) = \beta + f_1(X_1) + f_2(X_2) + \ldots + f_p(X_p), \qquad (3.17)$$

where Y is the response, $X_1, \ldots, X_p$ are the predictors, and $f_1, f_2, \ldots, f_p$ are some unknown smooth functions with several choices. For example, f can be

- a univariate smooth function fitted by some nonlinear smoother to model the main effect of a numeric variable;

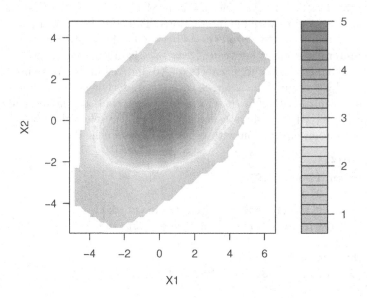

FIGURE 3.9
The partial dependence plot for $X3$ and $X4$.

- a bivariate smooth function to model the interactive effects between two numeric variables;

- a step function to model the main effect of a categorical input variable;

- a set of smooth functions to model the interactive effects between a categorical and a numeric variables.

Note that users can enforce some terms in the GAM to have parametric forms such as linear models.

On one hand, the GAM is more flexible than the traditional linear models, whose functional forms are often too rigid to accommodate nonlinear and interactive relationships among the response and input variables. On the other hand, GAM is simpler and more interpretable than many machine learning methods (e.g. artificial neural networks), which are usually considered as the *black-box* algorithms with little interpretations and insights of the fitted model. Like generalized linear model (GLM), GAM can be extended to various types of response variable

$$g[E(Y)] = \beta + f_1(X_1) + f_2(X_2) + \ldots + f_p(X_p), \qquad (3.18)$$

where $g(\cdot)$ is the link function. For regression, logistic regression and Poisson regression, $g(\cdot)$ are the identity link, logit and logarithm, respectively.

3.4.2 Smoothing Spline

A few nonlinear smoothers are available in GAM. One of the popular smoothers is smoothing spline, which is defined to be the function that minimize

$$\sum_{i=1}^{n}(y_i - \hat{f}(x_i))^2 + \lambda \int [f''(x)]^2 dx, \qquad (3.19)$$

where f'' is the second-order derivative of f over the class of all second-order differentiable functions. The smoothing parameter λ is a non-negative value which balances the goodness of fit of the data (i.e. the SSE part $\sum_{i=1}^{n}(y_i - \hat{f}(x_i))^2$) and the roughness of the function estimate (i.e. $\int [f''(x)]^2 dx$). If $\lambda = 0$, then the fitted smoothing splines pass through all the data points that have unique values on predictors x_i. On the other hand, if $\lambda = \infty$, then the fitted smoothing spline is the simple linear least square fit. The tuning parameter λ controls the roughness of the function estimate.

Although the minimizer of Eq. 3.19 is defined on an infinite-dimensional function space (i.e. Sobolev space), the solution is explicit, of finite-dimensional and unique. Based on the Representer Theorem [44], the minimizer of Eq. 3.19 is a weighted sum of natural cubic splines

$$f(x) = \sum_{i=1}^{n} w_i N_i(x), \qquad (3.20)$$

where $\{N_i(x)\}_{i=1}^{n}$ are a set of natural cubic spline basis functions with knot at each of the unique values of x. The weights w_i can be solved as a generalized ridge regression problem. The optimal value of λ can be chosen through cross-validation or *generalized cross-validation* (GCV), which is a computationally efficient approach to approximate the *leave-one-out cross-validation* (LOOCV) error.

The upper panel of Figure 3.10 shows a set of natural cubic spline basis (i.e. $N_i(x)$). The lower panel shows the fitted solid curve, which is the weighted sum of natural cubic splines basis functions (i.e. $w_i N_i(x)$). The height of the peak for each basis function is proportional to its weight w_i, which can be negative.

3.4.3 Revisit the Simulation Example

Let's revisit the simulation example described in Section 3.3.4. Currently, there are a few R packages that can fit GAM models, for example gam, mgcv, and gss. The mgcv package, which can automatically select the optimal value

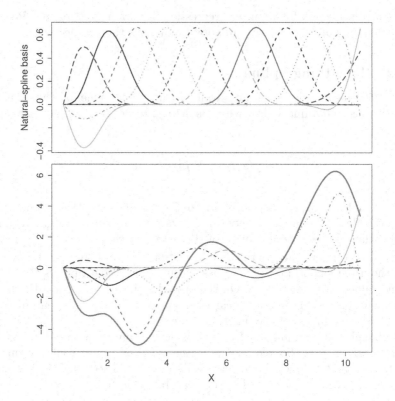

FIGURE 3.10
Upper: natural cubic spline basis functions. Lower: fitted function (red) from
the weighted natural cubic spline basis.

of the tuning parameter (i.e. λ in smoothing spline) for each term, is used
here. By default, the optimal value of λ is chosen to minimize the GCV error
in `mgcv` package.

GAM `fit1` fits an additive model with only main effects for $X_1 \sim X_4$.
Suppose we suspect there is an interaction between $X3$ and $X4$. We can fit
another GAM model `fit2` with the interaction, and compare to `fit1`. The `R`
code and partial outputs are shown below.

```
> library(mgcv)
> fit1=gam(y~s(X1)+s(X2)+s(X3)+s(X4),data=data)
> fit2=gam(y~s(X1)+s(X2)+s(X3,X4),data=data)
> summary(fit1)

Family: gaussian
Link function: identity

Formula:
```

```
y ~ s(X1) + s(X2) + s(X3) + s(X4)
```

```
Approximate significance of smooth terms:
       edf Ref.df      F p-value
s(X1) 5.665  6.872 150.10  <2e-16 ***
s(X2) 5.907  7.106 171.24  <2e-16 ***
s(X3) 5.635  6.861  92.31  <2e-16 ***
s(X4) 7.304  8.327 197.41  <2e-16 ***
```

```
R-sq.(adj) =  0.841   Deviance explained = 84.5%
GCV = 0.48133  Scale est. = 0.46905   n = 1000
> summary(fit2)
Family: gaussian
Link function: identity
```

```
Formula:
y ~ s(X1) + s(X2) + s(X3, X4)
```

```
Approximate significance of smooth terms:
          edf Ref.df       F p-value
s(X1)    2.148  2.762 733.6  <2e-16 ***
s(X2)    6.167  7.358 275.6  <2e-16 ***
s(X3,X4) 23.011 27.020 251.3  <2e-16 ***
```

```
R-sq.(adj) =  0.921   Deviance explained = 92.4%
GCV = 0.23986  Scale est. = 0.23211   n = 1000
```

From the models' summary, we see that both models have highly significant effects for all the smooth terms. However, the main effect model (i.e. `fit1`) only explains 84.5% of the deviance, while including the bivariate interaction between $X3$ and $X4$ explains an extra 7.9% (i.e. 92.4%-84.5%) of the deviance. We can formally test the significance of the bivariate interaction effect using the F test in `anova` function.

```
> anova(fit1,fit2,test="F")
Analysis of Deviance Table
```

```
Model 1: y ~ s(X1) + s(X2) + s(X3) + s(X4)
Model 2: y ~ s(X1) + s(X2) + s(X3, X4)
  Resid. Df Resid. Dev    Df Deviance      F   Pr(>F)
1    969.83     457.09
2    961.86     224.60 7.9735   232.48 125.62 < 2.2e-16 ***
```

From the outputs, we see that by including the bivariate interaction between X_3 and X_4, the deviance is reduced by 232.48 with additional 8 degrees

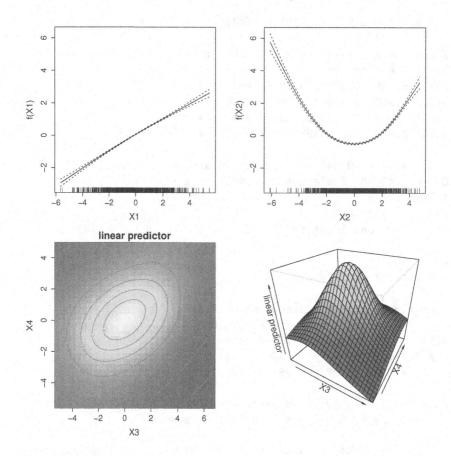

FIGURE 3.11
The component plots for $X1$ (upper left), $X2$ (upper right), contour plot (lower left) and 3D perspective plot (lower right) for bivariate interaction between $X3$ and $X4$.

of freedom. The p-value of the F-test indicates the interaction term is highly significant.

```
> par(mfrow=c(2,2))
> plot(fit2,select=1,ylab="f(X1)")
> plot(fit2,select=2,ylab="f(X2)")
> vis.gam(fit2,view=c("X3","X4"),plot.type="contour")
> vis.gam(fit2,view=c("X3","X4"),theta=30,phi=30,
  plot.type="persp")
```

Figure 3.11 shows the components plots from GAM `fit2`. We see the main effects for $X1$ and $X2$ are linear and quadratic, respectively, which agrees with the underlying true model defined in Eq. 3.16. The tick marks at the bottom of plots show the distribution of the corresponding variable in the data set. We can also graphically examine the bivariate contour plot (lower left) and 3D perspective plot (lower right) using the `vis.gam` function in the `mgcv` package. Both of them show that the interaction term has the peak around the origin.

4

The General Third-Variable Effect Analysis Method

In third-variable analysis, the important inference is to measure the direct effect of a variable X on the response variable Y and the indirect effect of X on Y through third-variable M. The third-variables can take the form as a mediator or as a confounder. When there are more than one third-variables, we should be able to differentiate the indirect effect from each of them. For the objective, we first propose general definitions of third-variable effects in the counterfactual framework. Figure 4.1 is a graphical model showing relationships among the exposure/independent variable X, the response variable Y, third variables M (M can be a vector), and other covariates $\mathbf{Z}$.

In Section 4.1, we present the notations used in the chapter. Section 4.2 proposes general definitions of third-variable effects. Section 4.3 shows some results of the third-variable effect analysis based on linear regression and logistic regression for different types of third-variables, which includes the estimates of third-variable effects and their variances through the Delta method. In Section 4.4, we propose algorithms for the third-variable analysis with non/semi-parametric predictive models and binary exposures. The algorithm is further developed to deal with continuous exposure variables in Section 4.5. The R package, SAS Macros and codes to perform the analysis is discussed in Chapter 5. For the purpose of understanding the concepts, using the R package and SAS macros to perform analysis, and interpreting results, readers can skip Sections 4.3 to 4.5.

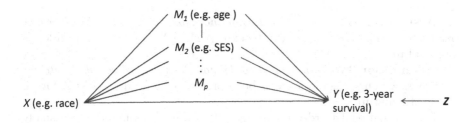

FIGURE 4.1
Interaction plot for independent variable X, mediators M_j, response Y and other explanatory variables $\mathbf{Z}$.

DOI: 10.1201/9780429346941-4

This chapter is derived in part from an article published in the Journal of Biometrics & Biostatistics [79], available online: https://www.hilarispublisher .com/open-access/general-multiple-mediation-analysis-with-an-application-to-expl-ore-racial-disparities-in-breast-cancer-survival-2155-6180.1000189.pdf.

4.1   Notations

Let $\mathbf{M} = (M_1, \cdots, M_p)^T$ be the vector of third-variables, where M_j be the *jth* mediator or confounder. $\mathbf{Z}$ is the vector of other explanatory variables that directly relate to the outcome Y, but is independent with respect to the exposure variable X. We are interested in exploring the mechanism of the change in the response, Y, when X varies. Let $Y(x, \mathbf{m})$ be the potential outcome if the exposure X is at the level of x, and $\mathbf{M}$ at $\mathbf{m}$. Denote the domain of X as dom_X. Let u^* be the infimum positive unit such that there is a biggest subset of dom_X, denoted as dom_{X^*}, in which any $x \in dom_{X^*}$ implies that $x + u^* \in dom_X$, and that $dom_X = dom_{X^*} \cup \{x + u^* | x \in dom_{X^*}\}$. If u^* exists, it is unique. Note that if X is continuous, $u^* = 0^+$ and $dom_X = dom_{X^*}$; if X is a binary variable, taking the values 0 or 1, then $u^* = 1$ and $dom_{X^*} = \{0\}$.

For random variables A, B and C, $A \perp\!\!\!\perp B | C$ denotes that given C, A is conditionally independent of B. In addition, $E(x)$ denotes the expectation of the random variable x.

4.2   Definitions of Third-Variable Effects

In this section, we define the third-variable effects which include direct effect, indirect effect, total effect and relative effect. For the identification of third-variable effects, three assumptions are required, which were also stated in [76]:

Assumption 1 *No-unmeasured-confounder for the exposure–outcome relationship.* This assumption can be expressed as $Y(x, \mathbf{m}) \perp\!\!\!\perp X | \mathbf{Z}$ for all levels of x and $\mathbf{m}$.

Assumption 2 *No-unmeasured-confounder for the exposure–third-variable relationship.* This assumption can be expressed as $X \perp\!\!\!\perp \mathbf{M}(x) | \mathbf{Z}$ for all levels of x and $\mathbf{m}$.

Assumption 3 *No-unmeasured-confounder for the third-variable–outcome relationship.* This assumption can be expressed as $Y(x, \mathbf{m}) \perp\!\!\!\perp \mathbf{M} | \mathbf{Z}$ for all levels of x and $\mathbf{m}$.

For conventional third-variable analysis, a fourth assumption is usually required: **Assumption 4** *Any third-variable is not causally prior to other third-variables.* This assumption is not required in the general third-variable analysis proposed in this book. In Chapter 6, we discuss in detail the implication of each of the four assumptions and implement a series of sensitivity analyses when any assumption is violated.

4.2.1 Total Effect

We define the total effect in terms of the average changing rate of Y with the treatment/exposure variable, X.

Definition 4.1 *Given* $\mathbf{Z}$*, the* **total effect** *(TE) of* X *on* Y *at* $X = x^*$ *is defined as the changing rate in* $E(Y(X, \mathbf{Z}))$ *when* X *changes:* $TE_{|\mathbf{Z}}(x^*) = \lim_{u \to u^*} \frac{E[Y(x^*+u)-Y(x^*)|\mathbf{Z}]}{u}$*. The* **average total effect** *is defined as* $ATE_{|\mathbf{Z}} = E_{x^*}\left[TE_{|\mathbf{Z}}(x^*)\right]$*, where* f *is the density of* x *and the density of* x^* *is derived as* $f^*(x^*) = \frac{f(x^*)}{\int_{x \in dom_{x^*}} f(x)dx}$*.*

With the assumptions 1-3, it is easy to see that

$$TE_{|\mathbf{Z}}(x^*) = \lim_{u \to u^*} \frac{E(Y|\mathbf{Z}, X = x^* + u) - E(Y|\mathbf{Z}, X = x^*)}{u}.$$

Note that if X is binary, $dom_{x^*} = \{0\}$, so the (average) total effect is $E(Y|\mathbf{Z}, X = 1) - E(Y|\mathbf{Z}, X = 0)$, which is equivalent to the commonly used definition of total effect in literature for binary outcomes. When $E(Y|X)$ is differentiable in X, the total effect is defined as $TE_{|\mathbf{Z}}(x^*) = \frac{\partial E(Y|\mathbf{Z},X)}{\partial X}|_{X=x^*}$.

Compared with conventional definitions of the average total effect that look at the difference in expected Y when X changes from x to x^*, we define the total effect based on the rate of change. The motivation is that the effect does not change with either the unit or the changing amount, $(x^* - x)$, of X, thus generalizing the definitions of third-variable effects. The definition of third-variable effects is consistent for exposure variables measured at any scale (binary, multi-categorical, or continuous). In addition to the average effect, using the changing rate can also help depict third-variable effects at any value of X in the domain.

4.2.2 Direct and Indirect Effects

We define the direct effect by fixing M at its marginal distribution, $f(M|\mathbf{Z})$, even when the value of X changes. When there is only one third-variable M, the direct effect of X at x^* is defined as $DE_{|\mathbf{Z}}(x^*) = E_M\left[\lim_{u \to u^*} \frac{E(Y(x^*+u,M)-Y(x^*,M))|\mathbf{Z})}{u}\right]$, where the expectation is taken at the marginal distribution of M. When Assumptions 1 and 2 are satisfied, $DE_{|\mathbf{Z}}(x^*) = E_M\left[\lim_{u \to u^*} \frac{E(Y|X=x^*+u,M,\mathbf{Z})-E(Y|X=x^*,M,\mathbf{Z})}{u}\right]$, where m takes the values in the support set $\{m|Pr(m|X = x^*) > 0 \ \& \ Pr(m|X = x^* + u^*) > 0\}$. The average direct effect is defined as $E_{x^*}[DE_{|\mathbf{Z}}(x^*)]$. In the same vein, we extend the definition of direct effect to multiple third-variables.

Definition 4.2 *For given* $\mathbf{Z}$, *the* **direct effect** *(DE) of* X *on* Y *not from* M_j *is defined as* $DE_{\backslash M_j|\mathbf{Z}}(x^*) =$

$$E_{M_j}\left[\lim_{u\to u^*}\frac{E(Y(x^* + u, M_j, \mathbf{M}_{-j}(x^* + u))) - E(Y(x^*, M_j, \mathbf{M}_{-j}(x^*)))}{u}\middle|\mathbf{Z}\right],$$

and the **average direct effect** *of* X *on* Y *not from* M_j *is defined as* $ADE_{\backslash M_j|\mathbf{Z}} = E_{x^*}DE_{\backslash M_j|\mathbf{Z}}(x^*)$, *where* $\mathbf{M}_{-j}$ *denotes the vector* $\mathbf{M}$ *without* M_j.

Note that if there are overlapping pathways through multiple third-variables, Assumption 4 is violated. We recommend using the sensitivity analysis proposed by [39] to assess the robustness of empirical results. If causal relationships among third-variables exist, one solution for third-variable analysis is to combine the causally related third-variables together and find the group effect. For example, if the real relationship is $X \to M_{t1} \to M_{t2} \to Y$, we group M_{t1} and M_{t2}, and consider the pathway $X \to \binom{M_{t1}}{M_{t2}} \to Y$ instead, where the third-variable is a vector. The within $\binom{M_{t1}}{M_{t2}}$ effects can also be explored separately through sequential mediation analysis.

Under the Assumptions 1-3, the average direct effect from exposure to outcome not through M_j is defined as

$$E_{x^*}E_{m_j}\{\lim_{u\to u^*}[E_{\mathbf{M}_{-j}|X=x^*+u}[E(Y|\mathbf{Z}, M_j = m_j, \mathbf{M}_{-j}, X = x^* + u)]-$$

$$E_{\mathbf{M}_{-j}|X=x^*}[E(Y|\mathbf{Z}, M_j = m_j, \mathbf{M}_{-j}, X = x^*)]]/u\}.$$

It is easy to show that when X is binary,

$$\begin{aligned}ADE_{\backslash M_j|Z} &= DE_{\backslash M_j|Z}(0)\\ &= E_{m_j}\{E_{\mathbf{M}_{-j}|X=1}[E(Y|\mathbf{Z}, M_j = m_j, \mathbf{M}_{-j}, X = 1)] -\\ &\quad E_{\mathbf{M}_{-j}|X=0}[E(Y|\mathbf{Z}, M_j = m_j, \mathbf{M}_{-j}, X = 0)]\}.\end{aligned}$$

Mathematically, the definitions of TE and DE differ in that for given $\mathbf{Z}$, the former takes the conditional expectation over M_j for given x^* while the latter takes the marginal expectation over M_j. The direct effect not from M_j measures the changing rate in the potential outcome with X, where M_j is fixed at its marginal level, while all the other third-variables can change with X. As in Figure 4.1, DE not from M_j is calculated as TE except for that the line between X and M_j is broken. Therefore, we call the changing rate in Y the direct effect of X on Y not from M_j, which is the summation of the direct effect of X on Y and the indirect effects through third-variables other than M_j.

With definitions 4.1 and 4.2, the definition of indirect effect is straightforward.

Definition 4.3 *Given* $\mathbf{Z}$, *the* **indirect effect** *of* X *on* Y *through* M_j *is defined as* $IE_{M_j|\mathbf{Z}}(x^*) = TE_{|\mathbf{Z}}(x^*) - DE_{\backslash M_j|\mathbf{Z}}(x^*)$. *Similarly, the average indirect effect through* M_j *is* $AIE_{M_j|\mathbf{Z}} = ATE_{|\mathbf{Z}} - ADE_{\backslash M_j|\mathbf{Z}}$.

Finally, the average direct effect of X not from the group of third-variables $\mathbf{M}$ is defined as $ADE_{|\mathbf{Z}} = ATE_{|\mathbf{Z}} - AIE_{\mathbf{M}|\mathbf{Z}}$. When $\mathbf{M}$ includes all potential third-variables, $ADE_{|\mathbf{Z}}$ is the average direct effect of X on Y. Note that sometimes the direct effect $DE_{|\mathbf{Z}}(x) = TE_{|\mathbf{Z}}(x) - IE_{\mathbf{M}|\mathbf{Z}}(x)$ is more relevant than ADE when the direct effects from the exposure are not constant at different exposure levels.

4.2.3   Relative Effect

The relative effect is defined as the direct effect or indirect effect divided by the total effect. For example, the relative effect of M_j at $X = x$ is calculated as $RE_{M_j|\mathbf{Z}}(x) = \frac{IE_{M_j|\mathbf{Z}}(x)}{TE_{|\mathbf{Z}}(x)}$. The relative effect of M_j can be explained as the proportion of effect that can be changed if one can manipulate so that M_j is adjusted to be at its marginal distribution when X changes. The average relative effect is the average direct effect or average indirect effect divided by the average total effect. Note that relative effects can be negative. If $RE_{M_j|\mathbf{Z}}(x) < 0$, it indicates that when X changes around x, if M_j is manipulated to be set at its marginal distribution, the magnitude of the total effect would be enlarged rather than explained. We discuss about how to explain a negative relative effect in more detail in the example in Section 5.3.

4.3   Third-Variable Effect Analysis with Generalized Linear Models

Based on the definitions in Section 4.2, third-variable effect analyses can be generalized to different types of exposure, outcome or third-variables, whose relationships can be modeled by different predictive models. Moreover, indirect effect from individual third-variable can be differentiated from multiple variables. In this Section, we discuss the multiple-third-variable analysis that adopts generalized linear models to fit relationships among variables.

4.3.1   Multiple Third-Variable Analysis in Linear Regressions

We first show that for continuous third-variables and outcomes that are modeled with linear regressions, the proposed average indirect effects are identical to those measured by the CD and CP methods. In addition, we show that the assumption of none exposure-third variable interaction which is necessary for the controlled direct effect is not required for the measurements of third-variable effects that are generally defined here. For simplicity, we ignore the covariate(s) $\mathbf{Z}$ for the following discussion. Without loss of generality, assume that there are two third-variables M_1 and M_2, and that the true relationships among variables are that:

$$M_{1i} = a_{01} + a_1 X_i + \epsilon_{1i}; \tag{4.1}$$

$$M_{2i} = a_{02} + a_2 X_i + \epsilon_{2i}; \tag{4.2}$$

$$Y_i = b_0 + b_1 M_{1i} + b_2 M_{2i} + c X_i + \epsilon_{3i}; \tag{4.3}$$

where the error terms $(\epsilon_{1i}, \epsilon_{2i})$ have identical and independent (iid) bivariate normal distributions, $\begin{pmatrix} \epsilon_{1i} \\ \epsilon_{2i} \end{pmatrix} \sim N_2 \left(\begin{pmatrix} 0 \\ 0 \end{pmatrix}, \begin{pmatrix} \sigma_1^2 & \rho\sigma_1\sigma_2 \\ \rho\sigma_1\sigma_2 & \sigma_2^2 \end{pmatrix} \right)$, and are independent with ϵ_{3i}, which are iid with the normal distribution $N(0, \sigma_3^2)$, for $i = 1, \dots, n$. Following Definitions 4.1 and 4.3, we have Theorem 4.1 for the measurements of direct and indirect effects.

Theorem 4.1 *The average total effect of X on Y is $a_1 b_1 + a_2 b_2 + c$; the average indirect effect through M_1 is $a_1 b_1$; and that through M_2 is $a_2 b_2$. Of all the effect from X to Y, a $\frac{a_1 b_1}{a_1 b_1 + a_2 b_2 + c}$ fraction is indirectly explained through M_1; $\frac{a_2 b_2}{a_1 b_1 + a_2 b_2 + c}$ fraction is from M_2; and $\frac{c}{a_1 b_1 + a_2 b_2 + c}$ fraction is directly from X.*

Proof

$$
\begin{aligned}
E(Y|X = x^* + u) &= \int_{m_1} \int_{m_2} E(Y|X = x^* + u, M_1, M_2) \cdot \\
&\qquad f(M_1, M_2 | X = x^* + u) \, dm_2 \, dm_1 \\
&= \int_{m_1} \int_{m_2} (b_0 + b_1 M_1 + b_2 M_2 + c(x^* + u)) \cdot \\
&\qquad f(M_1, M_2 | X = x^* + u) \, dm_2 \, dm_1 \\
&= b_0 + c(x^* + u) + b_1 E(M_1 | X = x^* + u) \\
&\quad + b_2 E(M_2 | X = x^* + u) \\
&= b_0 + c(x^* + u) + b_1(a_{01} + a_1(x^* + u)) \\
&\quad + b_2(a_{02} + a_2(x^* + u)) \\
E(Y|X = x^*) &= b_0 + cx^* + b_1(a_{01} + a_1 x^*) + b_2(a_{02} + a_2 x^*) \\
\therefore ATE &= \lim_{u \to u^*} \frac{E_{x^*}\{E(Y|X = x^* + u) - E(Y|X = x^*)\}}{u} \\
&= c + a_1 b_1 + a_2 b_2
\end{aligned}
$$

$$
ADE_{/M_1} = \lim_{u \to u^*} E_{x^*} \left\{ \frac{E_{M_1}[E_{M_2|X=x^*+u}(E(Y|X = x^* + u, M_1, M_2))]}{u} \right.
$$
$$
\left. - \frac{E_{M_2|X=x^*}(E(Y|X = x^*, M_1, M_2))]}{u} \right\}
$$

$$= \lim_{u \to u^*} E_{x^*} \left\{ \frac{E_{M_1}[(b_0 + b_1 M_1 + b_2 E(M_2|X = x^* + u) + c(x^* + u))}{u} \right.$$

$$\left. - \frac{(b_0 + b_1 M_1 + b_2 E(M_2|X = x^*) + cx^*)]}{u} \right\}$$

$$= \lim_{u \to u^*} \frac{E_{x^*}\{b_2(E(M_2|X = x^* + u) - E(M_2|X = x^*)) + cu\}}{u}$$

$$= \lim_{u \to u^*} \frac{u(a_2 b_2 + c)}{u}$$

$$= a_2 b_2 + c$$

Similarly,

$$ADE_{/M_2} = a_1 b_1 + c$$

Therefore,

$$AIE_{M_1} = ATE - ADE_{/M_1} = a_1 b_1;$$

and

$$AIE_{M_2} = ATE - ADE_{/M_2} = a_2 b_2.$$

$$Q.E.D.$$

These measurements of average indirect effects are identical to those from the *CD* and *CP* methods in linear regressions. Note that correlations among third-variables are allowed here. Moreover, compared with the controlled direct effects, we do not require the none exposure-third-variable interaction assumption. To illustrate this, assume that there is an interaction effect of M_1 and X on Y, so that equation (4.3) becomes $Y_i = b_0 + b_1 M_{1i} + b_2 M_{2i} + cX_i + dX_i M_{1i} + \epsilon_{3i}$. Based on the definitions, we have the following Theorem for third-variable effects.

Theorem 4.2 *The total effect of X on Y at $X = x$ is $a_1 b_1 + a_2 b_2 + c + a_{01} d + 2a_1 dx$, among which the indirect effect through M_1 is $a_1 b_1 + a_1 dx$; and that through M_2 is $a_2 b_2$. The direct effect from X is $c + a_{01} d + a_1 dx$.*

When X has no effect on M_1 (*i.e.*, $a_1 = 0$) and consequently there is no indirect effect from M_1, the indirect effect from M_1 defined by Definition 4.3 is 0 by Theorem 4.2.

Similar to results from Theorem 4.2, the *natural direct effect* from X and the *natural indirect effect* of M_1 when X changes from x to x^* are $(c + a_{01} d + a_1 da^*)(x - x^*)$ and $(a_1 b_1 + a_1 da)(x - x^*)$ separately [76]. However, both the natural direct and natural indirect effects depend on the changing amount of X, $x - x^*$. Moreover, the natural direct effect depends further on the end value of X, x^*, while the natural indirect effect depends on the start value, x.

4.3.2 Multiple Third-Variable Analysis in Logistic Regression

In this section, we illustrate the general third-variable effect analysis when the third-variable M is not continuous and when the response variable is binary.

In such case, a logistic regression can be adapted to model the relationship between the exposure X and the third-variable M. The independent/exposure variable X can be continuous or categorical. In the following discussion, we assume that the exposure variable X is binary.

4.3.2.1   When M is Binary

We first consider the third-variable M_1 to be binary and assume the following underlying true models:

$$logit(Pr(M_{1i} = 1))   =   a_{01} + a_1 X_i;$$

$$\vdots$$

$$Y_i   =   b_0 + b_1 M_{1i} + \sum_{j=2}^{p} b_j M_{ji} + c X_i.$$

In this situation, the CP method cannot be used directly since a_1 denotes the changing rate of $logit(M_1 = 1)$ with X but not of M_1 directly, while b_1 is still the changing rate of Y_i with M_1. The CD method is not readily adaptable either. The reasons are that first, we have to assume two true models for Y, one fitted with both the exposure variable and all third-variables and the other with the exposure variable only. Second, the coefficients of X in two models fitted with different subsets of explanatory variables have different scales, and consequently are not comparable. This is because the logistic regression does not assume a random error term as does the linear regression. The variance of the error term is fixed in the logistic regression. When different sets of predictors are used in the logistic regression, instead of re-distributing variances to the error term, the coefficients for predictors are rescaled to ensure the variance of the error to be the constant. Third, using the CD method, the indirect effects from different third-variables are not separable. With the definitions in Section 4.2, we derive the neat results in Theorem 4.3 to measure the indirect effect of M_1.

Theorem 4.3 *Using the above notations, the indirect effect from X to Y through M_1 is $b_1 \left(\frac{e^{a_{01}+a_1}}{1+e^{a_{01}+a_1}} - \frac{e^{a_{01}}}{1+e^{a_{01}}} \right)$.*

Proof *Change the integration into summation in the proof of Lemma 1 to obtain TE and DE for binary third-variables.*

$$
\begin{aligned}
TE   &=   E(Y|X = 1) - E(Y|X = 0) \\
&=   [b_0 + c + b_1 E(M_1|X = 1) + \sum_{j=2}^{p} b_j E(M_j|X = 1)] \\
&\quad -[b_0 + b_1 E(M_1|X = 0) + \sum_{j=2}^{p} b_j E(M_j|X = 0)]
\end{aligned}
$$

$$= c + b_1 \cdot Pr(M_1 = 1|X = 1) + \sum_{j=2}^{p} b_j E(M_j|X = 1)$$

$$-b_1 \cdot Pr(M_1 = 1|X = 0) - \sum_{j=2}^{p} b_j E(M_j|X = 0)$$

$$= c + b_1 \left(\frac{e^{a_{01}+a_1}}{1 + e^{a_{01}+a_1}} - \frac{e^{a_{01}}}{1 + e^{a_{01}}} \right)$$

$$+ \sum_{j=2}^{p} b_j \left[E(M_j|X = 1) - E(M_j|X = 0) \right]$$

$$DE_{/M_1} = [b_0 + c + b_1 E(M_1) + \sum_{j=2}^{p} b_j E(M_j|X = 1)]$$

$$- [b_0 + b_1 E(M_1) + \sum_{j=2}^{p} b_j E(M_j|X = 0)]$$

$$= c + \sum_{j=2}^{p} b_j \left[E(M_j|X = 1) - E(M_j|X = 0) \right]$$

$$IE_{M_1} = TE - DE_{/M_1}$$

$$= b_1 \left(\frac{e^{a_{01}+a_1}}{1 + e^{a_{01}+a_1}} - \frac{e^{a_{01}}}{1 + e^{a_{01}}} \right)$$

Q.E.D.

Note that the indirect effect from M_1 is naturally separated into two parts: b_1, representing the effect of changes in M_1 on Y, and $\left(\frac{e^{a_{01}+a_1}}{1+e^{a_{01}+a_1}} - \frac{e^{a_{01}}}{1+e^{a_{01}}} \right)$, denoting the difference in the prevalence of $M_1 = 1$ when X changes from 0 to 1.

The counterfactual framework is popular in dealing with the special case when X is binary (0 or 1 denoting control or treatment), for example, the natural indirect effect, $\delta_i(x)$ (see Equation 2.2.1), and the natural direct effect, $\zeta_i(x)$ (see Equation 2.2.1), discussed in Chapter 2. Based on the definition, Imai et al. (2010a and 2010b) defined the average causal indirect effect (ACIE) as $\bar{\delta}(x) = E(\delta_i(x))$ and the direct effect as $\bar{\zeta}(x) = E(\zeta_i(x))$. Their methods have to make the assumption that there is no interaction between X and the indirect effect, i.e. $\bar{\delta}(0) = \bar{\delta}(1)$ or $\bar{\zeta}(0) = \bar{\zeta}(1)$. Otherwise, there could be two measurements of indirect effect or direct effect, which brings in challenges to generalizing the third-variable analysis to multi-categorical or continuous exposures. Our method relaxes this assumption. One can easily show that the average direct effect we defined for binary X in single third-variable scenario is $P(X = 0) \cdot \bar{\zeta}(0) + P(X = 1) \cdot \bar{\zeta}(1)$.

4.3.2.2   When M is Multi-categorical

When the third-variable is multi-categorical with $K + 1$ distinct categories, denote that M takes one of the values $0, 1, \cdots, K$, multinomial logit regression model can be adapted to fit the relationship between M and X. Assume the true models are that:

$$log \frac{Pr(M_i = 1)}{Pr(M_{1i} = 0)} = a_{01} + a_1 X_i;$$

$$\vdots$$

$$log \frac{Pr(M_i = K)}{Pr(M_{1i} = 0)} = a_{0K} + a_K X_i;$$

$$Y_i = b_0 + \sum_{k=1}^{K} b_{1k} I(M_i = k) + c X_i;$$

where $I(M = m)$ is 1 if $M = m$, and 0 otherwise. We have the following theorem for the indirect effect of M.

Theorem 4.4 *With the above assumptions and notations, the indirect effect from X to Y through M is $\sum_{k=1}^{K} b_{1k} \left(\frac{e^{a_{0k}+a_k}}{1+\sum_{j=1}^{K} e^{a_{0j}+a_j}} - \frac{e^{a_{0k}}}{1+\sum_{j=1}^{K} e^{a_{0j}}} \right).$*

Proof

$$E(Y|X) = b_0 + \sum_{k=1}^{K} b_{1k} E(I(M = k)) + cX$$

$$= b_0 + \sum_{k=1}^{K} b_{1k} Pr(M = k) + cX$$

$$= b_0 + cX + \frac{\sum_{k=1}^{K} b_k e^{a_{0k}+a_k X}}{1 + \sum_{j=1}^{K} e^{a_{0j}+a_j X}}$$

$$\therefore TE = E(Y|X = 1) - E(Y|X = 0)$$

$$= c + \sum_{k=1}^{K} b_k \left(\frac{e^{a_{0k}+a_k}}{1+\sum_{j=1}^{K} e^{a_{0j}+a_j}} - \frac{e^{a_{0k}}}{1+\sum_{j=1}^{K} e^{a_{0j}}} \right)$$

$$DE_{/M} = c$$

$$IE_M = \sum_{k=1}^{K} b_k \left(\frac{e^{a_{0k}+a_k}}{1+\sum_{j=1}^{K} e^{a_{0j}+a_j}} - \frac{e^{a_{0k}}}{1+\sum_{j=1}^{K} e^{a_{0j}}} \right)$$

$$Q.E.D.$$

Again the indirect effect from M is naturally separated into two parts: b_k, representing when M changes from the reference group ($M = 0$) to the

k group, the change in Y, and $\left(\frac{e^{a_{0k}+a_k}}{1+\sum_{j=1}^{K} e^{a_{0j}+a_j}} - \frac{e^{a_{0k}}}{1+\sum_{j=1}^{K} e^{a_{0j}}} \right)$, denoting the difference in the prevalence of $M = k$ when X changes from 0 to 1. The indirect effect of M is the summation of effect from the K groups compared with the reference group.

4.3.2.3 Delta Method to Estimate the Variances

The Delta method was built on the Taylor series to estimate variances of estimators that have an asymptotically normal distribution. For the univariate case, let $g(\cdot)$ be a function of $\cdot$ that is differentiable around a parameter θ. In addition, we have a sequence of random variables X_n that satisfies $\sqrt{n}[X_n - \theta]$ converges in distribution to $N(0, \sigma^2)$. By Taylor's theorem,

$$g(X_n) \approx g(\theta) + g'(\theta)(X_n - \theta).$$

By the assumption and Slutsky's theorem, we have that

$$\sqrt{n}[g(X_n) - g(\theta)] \xrightarrow{D} N(0, \sigma^2[g'(\theta)]^2),$$

where $\xrightarrow{D}$ means converges in distribution.

Extending to multivariate cases with consistent estimators, assume $\hat{\vec{\beta}}$ is a vector of estimators that converges in probability to its true value $\vec{\beta}$ and

$$\sqrt{n}[\hat{\vec{\beta}} - \vec{\beta}] \xrightarrow{D} N(\mathbf{0}, \Sigma).$$

Let h be a differentiable function on $\vec{\beta}$, the Delta method has that

$$\sqrt{n}[h(\hat{\vec{\beta}}) - h(\vec{\theta})] \xrightarrow{D} N(\vec{0}, \nabla h(\vec{\theta})^T \Sigma \nabla h(\vec{\theta})).$$

When the variable relationships are modeled with (generalized) linear models, Delta method can be used to estimate the variances of the third-variable effect estimators. In this section, we deduct the estimated variances when the third-variable is continuous, binary, or multi-categorical, for binary X (taking the value of 0 or 1, thus $AIE = IE$ for all types of third-variables). Assume M_1 is a continuous third-variable, M_2 is binary and M_3 is categorical with K categories. Further, we assume the underlying predictive models are:

$$M_{1i} = a_{01} + a_1 X_i + \epsilon_i;$$

$$logit(Pr(M_{2i})) = a_{02} + a_2 X_i;$$

$$log\frac{Pr(M_{3i} = 1)}{Pr(M_{3i} = 0)} = a_{031} + a_{31} X_i;$$

$$\vdots$$

$$log\frac{Pr(M_{3i} = K)}{Pr(M_{3i} = 0)} = a_{03K} + a_{3K} X_i;$$

$$Y_i = b_0 + b_1 M_{1i} + b_2 M_{2i} + \sum_{m=1}^{K} b_{3m} I_{(M_{3i}=m)} + cX_i.$$

According to the results in Theorems 4.1, 4.3 and 4.4, the estimated indirect effects transmitted by continuous third-variable M_1, binary third-variable M_2, and multi-categorical third-variable M_3 are $a_1 b_1$, $b_2 \left(\frac{e^{a_{02}+a_2}}{1+e^{a_{02}+a_2}} - \frac{e^{a_{02}}}{1+e^{a_{02}}} \right)$, and $\sum_{k=1}^{K} b_{3k} \left(\frac{e^{a_{03k}+a_{3k}}}{1+\sum_{j=1}^{K} e^{a_{03j}+a_{3j}}} - \frac{e^{a_{03k}}}{1+\sum_{j=1}^{K} e^{a_{03j}}} \right)$, respectively. Let $\hat{a}$ denotes the maximum likelihood estimator for the coefficient a. We have the estimates for the indirect effects:

$$\hat{IE}_{M_1} = \hat{a}_1 \hat{b}_1$$

$$\hat{IE}_{M_2} = \hat{b}_2 \left(\frac{e^{\hat{a}_{02}+\hat{a}_2}}{1 + e^{\hat{a}_{02}+\hat{a}_2}} - \frac{e^{\hat{a}_{02}}}{1 + e^{\hat{a}_{02}}} \right)$$

$$\hat{IE}_{M_3} = \sum_{k=1}^{K} \hat{b}_{3k} \left(\frac{e^{\hat{a}_{03k}+\hat{a}_{3k}}}{1 + \sum_{j=1}^{K} e^{\hat{a}_{03j}+\hat{a}_{3j}}} - \frac{e^{\hat{a}_{03k}}}{1 + \sum_{j=1}^{K} e^{\hat{a}_{03j}}} \right)$$

Theorem 4.5 *Using the above notations and the multivariate Delta method, the variances for the indirect effect estimates are that:*

$$Var(\hat{IE}_{M_1}) \approx \hat{b}_1^2 \sigma_{\hat{a}_1}^2 + \hat{a}_1^2 \sigma_{\hat{b}_1}^2,$$

$$Var(\hat{IE}_{M_2}) \approx \mathbf{P}' \mathbf{\Sigma}_{\hat{b}_2, \hat{a}_{02}, \hat{a}_2} \mathbf{P},$$

$$Var(\hat{IE}_{M_3}) \approx \mathbf{Q}' \mathbf{\Sigma}_{\hat{b}_3, \hat{a}_{03}, \hat{a}_3} \mathbf{Q},$$

where

$$\mathbf{P}' = \left(\frac{e^{\hat{a}_{02}}(e^{\hat{a}_2}-1)}{(1+e^{\hat{a}_{02}})(1+e^{\hat{a}_{02}+\hat{a}_2})} \quad \frac{\hat{b}_2(e^{\hat{a}_2}-1)e^{\hat{a}_{02}}(1-e^{2\hat{a}_{02}+\hat{a}_2})}{(1+e^{\hat{a}_{02}})^2(1+e^{\hat{a}_{02}+\hat{a}_2})^2} \quad \frac{e^{\hat{a}_{02}+\hat{a}_2}}{(1+e^{\hat{a}_{02}+\hat{a}_2})^2} \cdot \hat{b} \right),$$

$$\mathbf{\Sigma}_{\hat{b}_2, \hat{a}_{02}, \hat{a}_2} = \begin{pmatrix} \sigma_{\hat{b}_2}^2 & 0 & 0 \\ 0 & \sigma_{\hat{a}_{02}}^2 & \sigma_{\hat{a}_{02}\hat{a}_2} \\ 0 & \sigma_{\hat{a}_{02}\hat{a}_2} & \sigma_{\hat{a}_2}^2 \end{pmatrix},$$

$$\mathbf{Q}' = \left(\mathbf{Q}_1' \quad \mathbf{Q}_2' \quad \mathbf{Q}_3' \right),$$

$$\mathbf{\Sigma}_{\hat{b}_3, \hat{a}_{03}, \hat{a}_3} = \begin{pmatrix} \Sigma_{\hat{b}_3} & 0 & 0 \\ 0 & Diag(\sigma_{\hat{a}_{031}}^2 \cdots \sigma_{\hat{a}_{03K}}^2) & Diag(\sigma_{\hat{a}_{031}\hat{a}_{31}} \cdots \sigma_{\hat{a}_{03K}\hat{a}_{3K}}) \\ 0 & Diag(\sigma_{\hat{a}_{031}\hat{a}_{31}} \cdots \sigma_{\hat{a}_{03K}\hat{a}_{3K}}) & Diag(\sigma_{\hat{a}_{31}}^2 \cdots \sigma_{\hat{a}_{3K}}^2) \end{pmatrix},$$

the mth element of $1 \times K$ vectors $\mathbf{Q}_1'$, $\mathbf{Q}_2'$ and $\mathbf{Q}_3'$ is:

$$(\mathbf{Q}_1')_m = \frac{e^{\hat{a}_{03m}+\hat{a}_{3m}}}{1 + \sum_{j=1}^{K} e^{\hat{a}_{03j}+\hat{a}_{3j}}} - \frac{e^{\hat{a}_{03m}}}{1 + \sum_{j=1}^{K} e^{\hat{a}_{03j}}},$$

$$(\mathbf{Q}_2')_m = \hat{b}_{3m} \cdot e^{\hat{a}_{03m}} \cdot \left(\frac{e^{\hat{a}_{03m}}(1 + \sum_{j=1, j\neq m}^{K} e^{\hat{a}_{03j}+\hat{a}_{3j}})}{(1 + \sum_{j=1}^{K} e^{\hat{a}_{03j}+\hat{a}_{3j}})^2} - \frac{1 + \sum_{j=1, j\neq m}^{K} e^{\hat{a}_{03j}}}{(1 + \sum_{j=1}^{K} e^{\hat{a}_{03j}})^2} \right),$$

$$(\mathbf{Q'_3})_m = \hat{b}_{3m} \cdot e^{\hat{a}_{03m}+\hat{a}_{3m}} \cdot \frac{1+\sum_{j=1,j\neq m}^{K} e^{\hat{a}_{03j}+\hat{a}_{3j}}}{(1+\sum_{j=1}^{K} e^{\hat{a}_{03j}+\hat{a}_{3j}})^2},$$

and matrix $\Sigma_{\hat{b}_3}$ is the variance-covariance matrix for the vector of estimators $(\hat{b}_{31}\cdots\hat{b}_{3K})$.

Proof

Let $f_1 = I\hat{E}_{M_1} = \hat{a}_1\hat{b}_1$, $f_2 = I\hat{E}_{M_2} = \hat{b}_2\left(\frac{e^{\hat{a}_{02}+\hat{a}_2}}{1+e^{\hat{a}_{02}+\hat{a}_2}} - \frac{e^{\hat{a}_{02}}}{1+e^{\hat{a}_{02}}}\right)$, *and* $f_3 = I\hat{E}_{M_3} = \sum_{m=1}^{K} \hat{b}_{3m}\left(\frac{e^{\hat{a}_{03m}+\hat{a}_{3m}}}{1+\sum_{j=1}^{K} e^{\hat{a}_{03j}+\hat{a}_{3j}}} - \frac{e^{\hat{a}_{03m}}}{1+\sum_{j=1}^{K} e^{\hat{a}_{03j}}}\right)$.

$$Var(I\hat{E}_{M_1}) \approx \begin{pmatrix} \frac{\partial f_1}{\partial \hat{a}_1} & \frac{\partial f_1}{\partial \hat{b}_1} \end{pmatrix} \cdot \begin{pmatrix} \sigma_{\hat{a}_1}^2 & 0 \\ 0 & \sigma_{\hat{b}_1}^2 \end{pmatrix} \cdot \begin{pmatrix} \frac{\partial f_1}{\partial \hat{a}_1} & \frac{\partial f_1}{\partial \hat{b}_1} \end{pmatrix}' = \hat{b}_1^2\sigma_{\hat{a}_1}^2 + \hat{a}_1^2\sigma_{\hat{b}_1}^2$$

$$Var(I\hat{E}_{M_2}) \approx \begin{pmatrix} \frac{\partial f_2}{\partial \hat{b}_2} & \frac{\partial f_2}{\partial \hat{a}_{02}} & \frac{\partial f_2}{\partial \hat{a}_2} \end{pmatrix} \cdot \begin{pmatrix} \sigma_{\hat{b}_2}^2 & 0 & 0 \\ 0 & \sigma_{\hat{a}_{02}}^2 & \sigma_{\hat{a}_{02}\hat{a}_2} \\ 0 & \sigma_{\hat{a}_{02}\hat{a}_2} & \sigma_{\hat{a}_2}^2 \end{pmatrix} \cdot \begin{pmatrix} \frac{\partial f_2}{\partial \hat{b}_2} & \frac{\partial f_2}{\partial \hat{a}_{02}} & \frac{\partial f_2}{\partial \hat{a}_2} \end{pmatrix}'$$

$$\frac{\partial f_2}{\partial \hat{b}_2} = \frac{e^{\hat{a}_{02}}(e^{\hat{a}_2}-1)}{(1+e^{\hat{a}_{02}})(1+e^{\hat{a}_{02}+\hat{a}_2})}$$

$$\frac{\partial f_2}{\partial \hat{a}_{02}} = \hat{b}_2(e^{\hat{a}_2}-1)\cdot\frac{\partial}{\partial \hat{a}_{02}}\left(\frac{e^{\hat{a}_{02}}}{(1+e^{\hat{a}_{02}})(1+e^{\hat{a}_{02}}+\hat{a}_2)}\right)$$

$$= \frac{\hat{b}_2(e^{\hat{a}_2}-1)e^{\hat{a}_{02}}(1-e^{2\hat{a}_{02}+\hat{a}_2})}{(1+e^{\hat{a}_{02}})^2(1+e^{\hat{a}_{02}+\hat{a}_2})^2}$$

$$\frac{\partial f_2}{\partial \hat{a}_2} = \frac{\hat{b}_2 e^{\hat{a}_{02}}}{1+e^{\hat{a}_{02}}}\cdot\frac{\partial}{\partial \hat{a}_2}\left(\frac{e^{\hat{a}_2}-1}{1+e^{\hat{a}_{02}+\hat{a}_2}}\right) = \frac{e^{\hat{a}_{02}+\hat{a}_2}}{(1+e^{\hat{a}_{02}+\hat{a}_2})^2}\cdot\hat{b}_2$$

Let $\mathbf{P'}=\left(\frac{e^{\hat{a}_{02}}(e^{\hat{a}_2}-1)}{(1+e^{\hat{a}_{02}})(1+e^{\hat{a}_{02}+\hat{a}_2})},\ \frac{\hat{b}_2(e^{\hat{a}_2}-1)e^{\hat{a}_{02}}(1-e^{2\hat{a}_{02}+\hat{a}_2})}{(1+e^{\hat{a}_{02}})^2(1+e^{\hat{a}_{02}+\hat{a}_2})^2},\ \frac{e^{\hat{a}_{02}+\hat{a}_2}}{(1+e^{\hat{a}_{02}+\hat{a}_2})^2}\cdot\hat{b}\right)$ *and*

$$\Sigma_{\hat{b}_2,\hat{a}_{02},\hat{a}_2} = \begin{pmatrix} \sigma_{\hat{b}_2}^2 & 0 & 0 \\ 0 & \sigma_{\hat{a}_{02}}^2 & \sigma_{\hat{a}_{02}\hat{a}_2} \\ 0 & \sigma_{\hat{a}_{02}\hat{a}_2} & \sigma_{\hat{a}_2}^2 \end{pmatrix},\ \text{then}\ Var(I\hat{E}_{M_2}) \approx \mathbf{P'}\Sigma_{\hat{b}_2,\hat{a}_{02},\hat{a}_2}\mathbf{P}.$$

$$Var(I\hat{E}_{M_3}) \approx \left(\left(\frac{\partial f_3}{\partial \mathbf{b}_3}\right)' \quad \left(\frac{\partial f_2}{\partial \hat{a}_{03}}\right)' \quad \left(\frac{\partial f_2}{\partial \hat{a}_3}\right)'\right)$$

$$\times \begin{pmatrix} \Sigma_{\hat{b}_3} & 0 & 0 \\ 0 & Diag(\sigma_{\hat{a}_{031}}^2\cdots\sigma_{\hat{a}_{03K}}^2) & Diag(\sigma_{\hat{a}_{031}\hat{a}_{31}}\cdots\sigma_{\hat{a}_{03K}\hat{a}_{3K}}) \\ 0 & Diag(\sigma_{\hat{a}_{031}\hat{a}_{31}}\cdots\sigma_{\hat{a}_{03K}\hat{a}_{3K}}) & Diag(\sigma_{\hat{a}_{31}}^2\cdots\sigma_{\hat{a}_{3K}}^2) \end{pmatrix}$$

$$\times \left(\left(\frac{\partial f_3}{\partial \mathbf{b}_3}\right)' \quad \left(\frac{\partial f_2}{\partial \hat{a}_{03}}\right)' \quad \left(\frac{\partial f_2}{\partial \hat{a}_3}\right)'\right)'$$

Let

$$\mathbf{Q'_1} = \left(\frac{\partial f_3}{\partial \hat{\mathbf{b}}_3}\right)' = \left(\frac{\partial f_3}{\partial \hat{b}_{31}} \cdots \frac{\partial f_3}{\partial \hat{b}_{3K}}\right)$$

where

$$\frac{\partial f_3}{\partial \hat{b}_{3m}} = \frac{e^{\hat{a}_{03m}+\hat{a}_{3m}}}{1+\sum_{j=1}^{K} e^{\hat{a}_{03j}+\hat{a}_{3j}}} - \frac{e^{\hat{a}_{03m}}}{1+\sum_{j=1}^{K} e^{\hat{a}_{03j}}}, m = 1, 2, \cdots, K;$$

Let

$$\mathbf{Q}_2' = \left(\frac{\partial f_3}{\partial \hat{\mathbf{a}}_{03}}\right)' = \left(\frac{\partial f_3}{\partial \hat{a}_{031}} \cdots \frac{\partial f_3}{\partial \hat{a}_{03K}}\right)$$

where

$$\frac{\partial f_3}{\partial \hat{a}_{03m}} = \hat{b}_{3m} \cdot e^{\hat{a}_{03m}} \cdot \left(\frac{e^{\hat{a}_{3m}}(1+\sum_{j=1,j\neq m}^{K} e^{\hat{a}_{03j}+\hat{a}_{3j}})}{(1+\sum_{j=1}^{K} e^{\hat{a}_{03j}+\hat{a}_{3j}})^2} - \frac{1+\sum_{j=1,j\neq m}^{K} e^{\hat{a}_{03j}}}{(1+\sum_{j=1}^{K} e^{\hat{a}_{03j}})^2}\right),$$

Let $\mathbf{Q}_3' = \left(\frac{\partial f_3}{\partial \hat{\mathbf{a}}_3}\right)' = \left(\frac{\partial f_3}{\partial \hat{a}_{31}} \cdots \frac{\partial f_3}{\partial \hat{a}_{3K}}\right)$ *where*

$$\frac{\partial f_3}{\partial \hat{a}_{3m}} = \hat{b}_{3m} \cdot e^{\hat{a}_{03m}+\hat{a}_{3m}} \cdot \frac{1+\sum_{j=1,j\neq m}^{K} e^{\hat{a}_{03j}+\hat{a}_{3j}}}{(1+\sum_{j=1}^{K} e^{\hat{a}_{03j}+\hat{a}_{3j}})^2}, m = 1, 2, \cdots, K.$$

Therefore, $Var(\hat{IE}_{M_3}) \approx \mathbf{Q}'\mathbf{\Sigma}_{\hat{b}_3,\hat{a}_{03},\hat{a}_3}\mathbf{Q}$, *where* $\mathbf{Q}' = \begin{pmatrix}\mathbf{Q}_1' & \mathbf{Q}_2' & \mathbf{Q}_3'\end{pmatrix}$ *and*

$$\mathbf{\Sigma}_{\hat{b}_3,\hat{a}_{03},\hat{a}_3} = \begin{pmatrix} \mathbf{\Sigma}_{\hat{b}_3} & \mathbf{0} & \mathbf{0} \\ \mathbf{0} & Diag(\sigma^2_{\hat{a}_{031}} \cdots \sigma^2_{\hat{a}_{3K}}) & Diag(\sigma_{\hat{a}_{031}\hat{a}_{31}} \cdots \sigma_{\hat{a}_{03K}\hat{a}_{3K}}) \\ \mathbf{0} & Diag(\sigma_{\hat{a}_{031}\hat{a}_{31}} \cdots \sigma_{\hat{a}_{03K}\hat{a}_{3K}}) & Diag(\sigma^2_{\hat{a}_{31}} \cdots \sigma^2_{\hat{a}_{3K}}) \end{pmatrix}.$$

Q.E.D.

The Delta method can be used to estimate variances of third-variable effects under more complicated cases. However, the deduction also becomes more complicated. In addition, when estimates cannot be written as functions of estimators that have asymptotic normal distributions, the Delta method cannot be used directly. For those cases, Bootstrap method can be used for the estimation of uncertainties.

4.4　Algorithms of Third-Variable Effect Analysis with General Predictive Models for Binary Exposure

When (generalized) linear regressions are insufficient in describing variable relationships, the third-variable analysis can be be very difficult. The following algorithms that derived directly from the general definitions of third-variable effects provide a non/semi-parametric method to calculate third-variable effects when the exposure variable is binary. With the algorithm, any predictive models can be used to model the relationship between the outcome and other variables. For the algorithm, we temporarily ignore other covariates $\mathbf{Z}$.

Algorithm 1 *Estimate the total effect: the total effect for binary exposure variable X is $E(Y|X = 1) - E(Y|X = 0)$. Under certain conditions, it can be directly obtained by averaging the response variable Y in subgroups of $X = 0$ and 1 separately and taking the difference.*

Algorithm 2 *Estimate the direct effect not through M_j, which is defined as $E_{m_j}\{E_{\mathbf{M}_{-j}|X=1}[E(Y|\mathbf{Z}, M_j = m_j, \mathbf{M}_{-j}, X = 1)] - E_{\mathbf{M}_{-j}|X=0}[E(Y|\mathbf{Z}, M_j = m_j, \mathbf{M}_{-j}, X = 0)]\}$:*

   1. Divide the original data sets $D_{-j} = \{\mathbf{M}_{-j,i}\}_{i=1}^n$,and $D_j = \{M_{ji}\}_{i=1}^n$ according to $X = 0$ or 1, denote the separated data sets as $D_{0,-j}$, $D_{1,-j}$ $D_{0,j}$ and $D_{1,j}$.

   2. Fit predictive model of Y on X and $\mathbf{M}$ on the whole data set, denote the predictive model as $E(Y) = f(X, M_j, \mathbf{M}_{-j})$.

   3. For $l = 1, \ldots, N$, where N is the number of iterations:

   (a) Sample $\left[\frac{n}{2}\right]$ M_js with replacement from $D_{0,j}$ and $D_{1,j}$ separately, randomly mix the two sets of samples to form a sample of M_j from its marginal distribution, which is denoted as $\{\tilde{M}_{jil}\}_{i=1}^n$.

   (b) Take $\left[\frac{n}{2}\right]$ samples with replacement from $D_{0,-j}$ and denote it as $\check{D}_{0,-j,l} = \{\check{\mathbf{M}}_{-j,i,l}\}_{i=1}^{\left[\frac{n}{2}\right]}$, where $[\cdot]$ denotes the largest integer that is smaller than $\cdot$. Take $\left[\frac{n}{2}\right]$ samples with replacement from $D_{1,-j}$ and denote it as $\check{D}_{1,-j,l} = \{\check{\mathbf{M}}_{-j,i,l}\}_{i=[\frac{n}{2}]+1}^n$. $\check{D}_{0,-j,l}$ and $\check{D}_{1,-j,l}$ are joint samples of $\mathbf{M}_{-j}$ from their conditional distributions on $X = 0$ or 1, respectively.

   (c) i. If M_j is categorical, taking $K + 1$ potential groups, let $\mu_{jkl0} = \text{average}\{f(X = 0, M_j = k, \check{\mathbf{M}}_{-j,i,l}), \text{for } i = 1, \ldots, \left[\frac{n}{2}\right] \text{ and } \tilde{M}_{jil} = k\}$ and $\mu_{jkl1} = \text{average}\{f(X = 1, M_j = k, \check{\mathbf{M}}_{-j,i,l}), \text{for } i = \left[\frac{n}{2}\right] + 1, \ldots, n, \text{ and } \tilde{M}_{jil} = k\}$, where $k = 0, \ldots, K$. Let $\widehat{DE}_{\backslash M_j,l} = \sum_{k=0}^K Pr(\tilde{M}_{jl} = k) \times (\mu_{jkl1} - \mu_{jkl0})$, where $Pr(\tilde{M}_{jl} = k)$ is the probability of $M_j = k$ in the sample $\{\tilde{M}_{jil}\}_{i=1}^n$.

   ii. If M_j is continuous, let $\widehat{DE}_{\backslash M_j,l} = \frac{1}{n - \left[\frac{2}{n}\right]} \sum_{i=\left[\frac{n}{2}\right]+1}^n$

   $f(1, \tilde{M}_{jil}, \check{\mathbf{M}}_{-j,i,l}) - \frac{1}{\left[\frac{n}{2}\right]} \sum_{i=1}^{\left[\frac{n}{2}\right]} f(0, \tilde{M}_{jil}, \check{\mathbf{M}}_{-j,i,l})$.

   4. The direct effect not from M_j is estimated as $\frac{1}{N} \sum_{l=1}^N \widehat{DE}_{\backslash M_j,l}$.

The model f fitted in step 2 can be parametric or nonparametric. In Section 5.3, we illustrate the method with both logistic model and multivariate additive regression trees (MART).

4.5 Algorithms of Third-Variable Effect Analysis with General Predictive Models for Continuous Exposure

When the exposure variable is continuous, we estimate the third-variable effect by using a numerical method for general predictive models. Let a has a positive value. Using the exposure variable to predict the third-variables, we assume the fitted models have the following form:

$$\left(\begin{array}{c} M_{1i} \\ M_{2i} \\ \ldots \\ M_{pi} \end{array} \middle| x_i \right) \sim \Pi\left(\left(\begin{array}{c} g_1(x_i) \\ g_2(x_i) \\ \ldots \\ g_p(x_i) \end{array} \right), \Sigma \right) \tag{4.4}$$

In addition, the predictive model for the response variable is assumed to be

$$Y_i = f(x_i, M_{1i}, \ldots, M_{pi}) + \epsilon_i, \tag{4.5}$$

where ϵ_i is the independent random error term and Π is the joint distribution of $\mathbf{M}$ given X, which has a mean vector $\mathbf{g}(x_i)$ and variance-covariance matrix Σ.

Assume we have the observations $(y_i, x_i, M_{1i}, \ldots, M_{pi})$, for $i = 1, \ldots, n$. Let $D_x = \{x_i | x_i \in domain_{x^*}\}$ and N be a large number. We have the following algorithms to estimate third-variable effects:

Algorithm 3 *Estimate the average total effect:*

    1. Draw N xs from D_x with replacement, denote that as $\{x_j, j = 1, \ldots, N\}$.

    2. Draw $(M_{1j1}, \ldots, M_{pj1})^T$ given $X = x_j$ from equation (4.4) for $j = 1, \ldots, N$.

    3. Draw $(M_{1j2}, \ldots, M_{pj2})^T$ given $X = x_j + a$ from equation (4.4) for $j = 1, \ldots, N$.

    4. The average total effect is estimated as $\widehat{ATE} = \frac{1}{Na}\left(\sum_{j=1}^{N} f(x_j + a, M_{1j2}, \ldots, M_{pj2}) - \sum_{j=1}^{N} f(x_j, M_{1j1}, \ldots, M_{pj1}) \right).$

Algorithm 4 *Estimate the average direct effect not through M_k:*

    1. Draw N M_ks from the observed $\{M_{ki}, i = 1, \ldots, n\}$, with replacement, denote as $M_{kj}, j = 1, \ldots, N$.

2. Draw $(M_{1j1}, \ldots, M_{k-1,j1}, M_{k+1,j1}, \ldots, M_{p_j1})^T$ *given* $X = x_j$ *from distribution derived from equation (4.4), where x_js were obtained in Algorithm 3, step 1, $j = 1, \ldots, N$.*

3. Draw $(M_{1j2}, \ldots, M_{k-1,j2}, M_{k+1,j2}, \ldots, M_{p_j2})^T$ *given* $X = x_j + a$ *from distribution derived from equation (4.4).*

4. The average direct effect not through M_k is estimated as
$$\widehat{ADE}_{\backslash M_k} = \frac{1}{Na} \left(\sum_{j=1}^{N} f(x_j + a, M_{1j2}, \ldots, M_{k-1,j2}, M_{kj}, \right.$$
$$\left. M_{k+1,j2}, \ldots, M_{pj2}) - \sum_{j=1}^{N} f(x_j, M_{1j1}, \ldots, M_{k-1,j1}, M_{kj}, \right.$$
$$\left. M_{k+1,j1}, \ldots, M_{pj1}) \right).$$

Note that the conditional distribution of $(M_{1j}, \ldots, M_{k-1,j}, M_{k+1,j}, \ldots, M_{p_j})^T$ given X can be easily derived from equation (4.4) if $\mathbf{\Pi}$ is multivariate normal. Then the marginal distribution of a subset of random variables is still multivariate normal with the mean vector and the covariance matrix deleting the rows/columns corresponding to the variables that has to be marginalized out.

The average indirect effect of M_k is estimated as $\widehat{AIE}_{M_k} = \widehat{ATE} - \widehat{ADE}_{\backslash M_k}$. We provide the algorithms for calculating the average third-variable effects. To calculate the third-variable effects at a certain value, say x_0, we just replace all x_js in above Algorithms with x_0.

5

The Implementation of General Third-Variable Effect Analysis Method

In Chapter 4, we discuss the general third-variable effect analysis method and the algorithms to implement the method. In this Chapter, we discuss the software, the R package *mma* and SAS macros, to implement the method. In addition, we give examples with real applications and simulations.

5.1   The R Package *mma*

We first discuss how to perform the general third-variable effect analysis with R. This section is derived in part from an article published in the Journal of Open Research Software [80], available online: https://openresearchsoftware.metajnl.com/articles/10.5334/ jors.160/.

R is a free statistical software. It can be downloaded from the website https://cran.r-project.org/ under many platforms. For basic knowledge about using R, readers are referred to the R Manuals provided in the cran website.

An R package, "mma", was compiled in the R environment to perform the third-variable effect analysis discussed in Chapter 4. To implement the method, we need to first install and call/load the package.

```
install.packages("mma")
library(mma)
```

Using the "mma" package, the third-variable effect analysis can be performed through three steps which are completed by three functions:

- Function *data.org* is used to identify potential third-variables and to prepare the data set for further analysis.

- Function *med* is used to implement algorithms for estimating third-variable effects based on the analytic data set formed by *data.org*.

- Function *boot.med* is for making inferences on third-variable effects using the bootstrap method.

On the other hand, the function *mma* executes all three steps with one command. In addition, we provide generic functions such as *print, summary* and *plot* to help present results and figures of analysis. All arguments used in these functions are fully documented within the package.

For illustration purposes, the package also includes a real data set *Weight_Behavior*. This data set was collected by Dr. Scribner from the Louisiana State University Health Sciences Center to explore the relationship between children weight (measured by a continuous variable *bmi* and a binary variable *overweight*) and behavior (such as snack, sports, computer hours, etc.) through a survey of children, teachers and parents in Grenada in 2014. This data set has 691 observations and 15 variables. Table 5.1 lists the descriptions of variables. We use the data set to illustrate the use of the R package *mma*. The illustration is based on the package *mma* version $8.0 - 0$ or later.

TABLE 5.1
Variables in the data set "Weight_Behavior".

Variable Name	Description
bmi	Body mass index
age	Age at survey
sex	Male or female
race	African American, Caucasian, Indian, Mixed or Other
numpeople	Number of people in family
car	Number of cars in family
gotosch	Four levels of methods to go to school
snack	Eat a snack in a day or not
tvhours	Number of hours watching TV per week
cmpthours	Number of hours using computer per week
cellhours	Number of hours playing with cell phones per week
sports	In a sport team or not
exercise	Number of hours of exercises per week
sweat	Number of hours of sweat-producing activities per week
overweigh	The child is overweight or not

5.1.1   Identification of Potential Mediators/Confounders and Organization of Data

The function *data.org* is used to identify important mediators/confounders (M/C). To be identified as an M/C, a variable typically satisfies two conditions. First, the variable is significantly correlated with the predictor. To test this, we use chi-square test, ANOVA, or the Pearson's correlation coefficient test, depending on the types (e.g. categorical, continuous, etc.) of the variable and the exposure variable(s). The significance level is set by the argument *alpha2* (0.1 by default). The second condition to identify a potential M/C is that the variable has to be significantly related with the outcome(s) adjusting (the argument is set as testtype=1, by default) or not adjusting (testtype=2) for the exposure and other variables. The significance level for this test is set by the argument *alpha* (0.1 by default). If both conditions are satisfied, the variable is included in further analysis as a potential M/C. If only the second condition is satisfied, the variable is included as a covariate (other explanatory variable, *z*, see Figure 4.1) but not an M/C, unless the variable is forced in as an important third-variable.

Note that the two tests are not the formal process to identify important third-variables. The formal test should be based on the final inferences on the estimated indirect effects. The testing steps in the *data.org* function are to provide summary and descriptive statistics about the linear relationships among variables. It also helps to pre-select variables as potential third-variables or covariates. Users can choose the significance levels according to their needs. In general, if the data set is small and in an effort to avoid leaving any important variable out, the significance level can be set at a little high, say 0.1. On the other hand, if the dataset is large and the purpose is to remove as many irrelevant variables as possible, Bonferroni correction can be applied to use small significance levels. When needed, variables can also be forced into analysis as either third-variables or covariates without tests, which is discussed later in this subsection.

The following codes are to identify mediator/confounders and covariates that explain the difference in being overweight by gender in the data set *weight_behavior*. All covariates and potential M/Cs are included in a data frame and read in as "x". The exposure variable (vector) is read in as "pred", and the response (vector) is read in as "y". In the example, the exposure is binary, *gender*, and the outcome, "y", is also a binary indicator of *overweight* (1) or not (0). The reference group of the exposure variable is *male* as is identified by the argument *predref= "M"*. There are 12 variables in *x*, which includes all potential M/Cs and covariates. The argument *mediator=5:12* indicates that the variables from column 5 to column 12 in *x* should be tested as potential M/Cs or covariates. Other columns in *x* that are not specified in *mediator* are forced in analysis as covariates. In the example, columns 1 to 4 in *x* are considered as covariates without the need of tests. The *mediator* argument can also be a character vector with names of the potential M/Cs in *x* (e.g. *mediator=c("age", "numpeople", ...))*.

```
data("weight_behavior")
#binary predictor x
#binary y
 x=weight_behavior[,c(2,4:14)]
 pred=weight_behavior[,"sex"]
 y=weight_behavior[,"overweigh"]
 data.b.b.1<-data.org(x,y,mediator=5:12,jointm=list(n=1,
    j1=7:9),pred=pred,predref="M", alpha=0.4,alpha2=0.4)
```

The argument *jointm* is used in the function to specify the group(s) of variables where the joint effect of all variables in the group is of interest. *jointm* is a list where the first element tells the total number of groups of interest, and each of the following items defines a group of variables to be considered jointly. A variable can show in more than one group. For each group defined by *jointm*, both the individual indirect effect of each variable and the joint indirect effect of all variables in the group are reported. If a variable is listed in the *jointm* argument, it is forced into the third-variable analysis as a potential third-variable without having to satisfy the two conditions. If a variable is included in *mediator* but not in *jointm*, the variable is to be tested for the two conditions listed above. If the second condition is not satisfied, the variable is dropped out for further analysis. If a researcher wants to check the individual third-variable effect of one or more variables even if those variables may not meet the two conditions, a trick is to form a group that include all those variables in *jointm*. In the above example, the joint effect of variables at columns 7, 8, and 9 of x is of interests. These variables are forced into analysis as third-variables. We set high significance levels (*alpha=0.4, alpha2=0.4*) to include more potential third-variables. This is not recommended in real practice.

The function data.org returns a *med_iden* object. The object is a list with two lists in it: *bin.results* and *cont.results*. If there is no binary or categorical exposure, *bin.results = NULL*. On the other hand, if there is no continuous exposure, *cont.results = NULL*. In the list *bin.results* and/or *cont.results*, x gives a data frame that includes all identified third-variables and covariates in explaining the outcome. Compared with the original x read in for *data.org* function, some variables are moved out of x through tests. *dirx* gives the data frame with all exposure variables; *contm, binm, catm* and *jointm* specify the positions of continuous, binary, categorical, and joint third-variables in the x.

The generic function *summary* helps summarize the *med_iden* object. It prints the test results and the lists of identified third-variables and covariates. The above code generates the object *data.b.b.1*, which is a *med_iden* object. *summary(data.b.b.1)* gives the following results:

```
>summary(data.b.b.1)
Identified as mediators:
[1] "tvhours"   "cmpthours" "cellhours" "exercises"
    "sweat"     "sports"
Selected as covariates:
[1] "age"       "race"      "numpeople" "car"
Tests:
            P-Value 1.y P-Value 2.predF
age             0.836              NA
race            0.850              NA
numpeople       0.561              NA
car             0.052              NA
gotosch         0.710              NA
```

snack	0.724	NA
tvhours -	0.830	0.484
cmpthours -	0.826	0.760
cellhours -	0.067	0.688
sports *	0.000	0.003
exercises *	0.176	0.203
sweat *	0.181	0.046
predF	0.178	NA

```
*:mediator,-:joint mediator
P-Value 1:Type-3 tests in the full model (data.org) or
estimated coefficients (data.org.big) when testtype=1,
univariate relationshiptest with the outcome when
testtype=2.
P-Value 2:Tests of relationship with the Predictor
```

In the results, the variables *tvhours, cmpthours, cellhours, exercises, sweat,* and *sports* are identified as third-variables, among which *tvhours, cmpthours,* and *cellhours* are forced in as third-variables since their joint effect is of interest. Selected M/Cs are indicated by a "*", forced-in M/Cs by a "-" following the variable names. The *P-Value 1.y* in the table gives the p-value of type-III test results for each covariate coefficient in predicting y if *testtype=1* (by default), or of the univariate relationship test between the outcome y and each covariate if *testtype=2*. If a variable identified in the *mediator* argument has a p-value 1 larger than the preset *alpha* and is not included in the *jointm* argument, the variable is not tested to get the second p-value. *P-Value 2.predF* gives the p-value of testing the relationship between the exposure variable, *predF*, and the variable named in the row.

An alternative way of specifying potential M/Cs rather than using the argument *mediator* is to use *binmed, catmed* and *contmed*. Using the general third-variable effect analysis method, the third-variables can be continuous or categorical. In *data.org* function, variables are identified as binary or categorical automatically if they are factors, characters, or have only 2 distinct values. Alternatively, a binary or categorical third-variable can be manually specified by listing it under the argument *binmed* (binary) or under *catmed* (categorical) but not under the argument *mediator*. By doing so, the reference group can be designated by the corresponding argument *binref* or *catref*. If not designated, the reference group for each categorical variable is the first level of the variable. The following is an example:

```
data.b.b.2<-data.org(x,y,pred=pred,contmed=c(7:9,11:12),
                binmed=c(6,10),binref=c(1,1),catmed=5,
                catref=1, predref="M",jointm=
                list(n=1,j1=7:9),alpha=0.4,alpha2=0.4)
summary(data.b.b.2)
```

In addition, to test the relationship between the exposure and potential

M/Cs, we can also adjust for other covariates. The other covariates for the exposure-M/C relationship are specified in the argument *cova*. If *cova* is a data frame, by itself, all covariates in cova are adjusted for all potential M/Cs. If *cova* is a list, the first item is the covariate data frame, and the second item list the names or column numbers of M/Cs in *x* that should be adjusted for the covariates.

Lastly, when fitting the relationship between the outcome and other variables, a generalized linear model is used. By default, the linkage function that links the mean of the outcome with the system component is *binomial(link = "logit")* for binary, or *gaussian(link= "identity")* for continuous outcome. The linkage function can also be otherwise specified in the *family1* argument. All families that can be used with the *glm* function are also available with the "mma" package.

Furthermore, observations can be assigned with different weights in the analysis, where weights of observations are input in the argument *w*.

Finally, both the exposure and outcome variables can be multivariate. In addition, the outcome variable can be a time-to-event variable so that we can deal with survival functions in third-variable analysis. We discuss these cases in other Chapters.

5.1.2   Third-Variable Effect Estimates

The *med* function is used to estimate the third-variable effect (TVE) using the output from the *data.org* function. The default approach is to use generalized linear modes to fit all models: the final full model(s) for predicting the outcome(s), and the predicting model for each third-variable. If in the *med* function, the argument *nonlinear* is set to be *TRUE* (*nonlinear=TRUE*), Multiple Additive Regression Trees (MART) are used to fit the relationship between outcome(s) and other variables. We use the R package *gbm* for the fitting of MART. Smoothing splines are used to fit the relationship between each potential M/C and exposure variables(s). We use the R package *splines* to generate the spline bases. Furthermore, users can self-define the final model for predicting the outcome(s) using the argument *custom.function*. Using the above example, the following codes are to perform a TVE analysis with a generalized linear model (*med.b.b.1*) or MART (*med.b.b.2*) on the *data.b.b.1* with identified third-variables and covariates.

```
med.b.b.1<-med(data=data.b.b.2,n=50)
med.b.b.2<-med(data=data.b.b.2,n=50,nonlinear=TRUE)
```

The n in codes is the N in Algorithms 2, 3 and 4, indicating the number of resampling times. By default, $n = 20$. For continuous outcomes, the argument *margin* defines the a in Algorithms 3 and 4, indicating the incremental amount in the exposure variable. By default, *margin* is 1. When generalized linear models are used for TVE analysis, the argument *family1* assigns the distribution(s) for the outcome(s), and the link functions that link the mean(s) of outcomes with the systematic components. As discussed before, by default, *gaussian(link= "identity")* is assigned to continuous outcome(s), and *binomial(link = "logit")* for binary outcome(s). If the outcome is a time-to-event variable, the *coxph* function in the "survival" package is used to fit the final model.

When *nonlinear=TRUE*, MART is used to fit the final full model. The *gbm* package is used for the model fitting, the interaction depth parameter is set by D (the argument *interaction.depth* in *gbm*), and the shrinkage parameter is set by *nu* (the argument *shrinkage* in *gbm*). The assumed distribution(s) of the outcome(s) is set by the argument *distn*. The default value of *distn* is *"gaussian"* for continuous outcome(s), *"bernoulli"* for binary and *"coxph"* for time-to-event outcome(s). Readers are referred to Chapter 3 for details about fitting MART. In addition, when *nonlinear=TRUE*, the relationships between the exposure variable(s) and each third-variable is fitted using natural cubic splines. The argument *df1* assigns the degree of freedom for splines. If the third-variable effects on a new set of exposures are of interest, users can set the new set by *pred.new* in the *med* function.

The *med* function generates an object with the *med* class. Two generic functions are compiled for the *med* class object to show results. One is the *print* function, to show the results of TVE analysis. The other is the *plot* function, to help visualize the third-variable effects and the relationships among variables. The following results are from the *print* function of the *med* object.

```
> med.b.b.1
For the predictor predF:
The estimated total effect:[1] 0.5695

The estimated indirect effect:
     y1.all    y1.tvhours y1.cmpthours y1.cellhours
    0.2208      -0.0196       0.0381        0.0295
   y1.exercises y1.sweat    y1.sports         y1.j1
   -0.0457  0.0033         0.1632        0.0563

> med.b.b.2
For the predictor predF :
The estimated total effect:[1] 0.09831

The estimated indirect effect:
     y1.all    y1.tvhours y1.cmpthours y1.cellhours
    0.0822       0.0025       0.0083        0.0057
   y1.exercises y1.sweat    y1.sports         y1.j1
    0.0029  -0.0004        0.0723        0.0182
```

For both models, the response variable is on the scale of log-odds of being overweight. The third-variable effects calculated by the logit model or MART are not exactly the same for this case. The main reason is that MART considers potential nonlinear relationships, while the logit regression model assumes a linear relationship between variables and the response. Also, note that the sum of direct and indirect effects may not equal to the total effects. This is because there are potential correlation and therefore overlapping third-variable effects among third-variables. If one would like to assume independent third-variable effects, calculating the total effect by adding up the direct and indirect effects is preferred.

Based on results from the non-linear modeling, overall, girls are more likely to be overweight than boys (the total effect is $0.09831 > 0$). That is on average, the odds of being overweight for girls is $exp(0.09831) = 1.1$ times that for boys. The joint third-variable effects of TV, computer, and cell phone hours on the sexual difference in the log-odds of being overweight is 0.0182, meaning that about $0.0182/0.09831 = 18.5\%$ of the sexual difference in overweight can be explained by the joint effects of weekly hours spent on TVs, computers, and/or cellphones. If both boys and girls spent about the same amounts of time on TV, computers, and cellphones at the population level, the sexual difference in overweight could be reduced by 18.5%. A potential explanation is that, on average, girls spend more time on TV, computers, and cellphones; and the amount of time spent on those activities is positively related with being overweight. The explanation may be visually shown by the *plot* function, we illustrate the plot results in the next subsection.

5.1.3   Statistical Inference on Third-Variable Effect Analysis

We use the bootstrap method to measure the uncertainty in estimating third-variable effects. The function *boot.med* is created for this purpose. The following codes are used to calculate the variances and confidence intervals of the estimated third-variable effects. The outputs returning from the *data.org* function is input as the first argument. The argument *weight* is a vector that assigns weights to observations when the observations are not treated equally in the analysis. For the bootstrap samples, the observations are drawn from the original data set with replacement and with probabilities proportional to the assigned weight. n_2 indicates the number of bootstrap samples. The third-variable effects, variances, and confidence intervals are estimated based on the estimates from bootstrap samples. The argument *nonlinear* is similarly defined as in the function *med*.

```
bootmed.b.b.1<-boot.med(data=data.b.b.2,n=2,n2=50)
bootmed.b.b.2<-boot.med(data=data.b.b.2,n=2,n2=40,
                nu=0.05,nonlinear=TRUE)
```

Note that for both functions *med* and *boot.med*, the results from the function *data.org* are input as an argument. However, to run the function *data.org* before running *med* and *boot.med* is not necessary. Users can set *data=FALSE*, and input all the needed arguments for *data.org* in *med* or *boot.med* to run these functions. The results from the *boot.med* function are classed as "mma".

Finally, the whole process, from identifying third-variables to estimating and making inferences on the TVEs, can be carried out by one function *mma*, as shown by the following codes. The function returns a "mma" class object, which is of the same format as the output of the *boot.med* function. If any manual intervention is required, separating functions are recommended.

```
x=weight_behavior[,c(2,4:14)]
pred=weight_behavior[,3]
y=weight_behavior[,15]
mma.b.b.glm<-mma(x,y,pred=pred,contmed=c(7:9,11:12),
           binmed=c(6,10),binref=c(1,1), catmed=5,
           catref=1,predref="M",alpha=0.4,
           alpha2=0.4, n=2,n2=50)
mma.b.b.mart<-mma(x,y,pred=pred,contmed=c(7:9,11:12),
           binmed=c(6,10),binref=c(1,1), catmed=5,
           catref=1,predref="M",alpha=0.4,alpha2=
           0.4, nonlinear=TRUE,n=2,n2=40)
```

Generic functions *print, summary*, and *plot* are created for the "mma" class to help users interpret the results easily. The *print* function for the "mma" class is similar to the *print* function for the "med" class. It prints the estimated third-variable effects only.

The *summary* function prints out more information: the estimation of third-variable effects by using the whole data set (under the notation *est*) or by averaging the estimates from bootstrap samples (estimates are noted as *mean*). It also calculates the estimated standard deviations (*sd*) and confidence intervals of the estimated third-variable effects from bootstrap samples. Three methods are used to build the confidence intervals: 1) normal approximation, noted as (*upbd, lwbd*), 2) quartiles of bootstrap estimates, noted as (*lwbd_q, upbd_q*) and 3) the ball estimates, noted as (*lwbd_b, upbd_b*). The ball estimates of intervals are to build confidence intervals with an overall significance level. The method to build the ball confidence intervals is discussed in Chapter 7.

```
> summary(bootmed.b.b.2,alpha=0.2)
MMA Analysis: Estimated Mediation Effects Using MART
For Predictor/Moderator at predF
$$total.effect
   est   mean     sd   upbd   lwbd upbd_q lwbd_q upbd_b lwbd_b
 0.170  0.284  0.164  0.494  0.074  0.508  0.112  0.584  0.022

$$direct.effect
   est   mean     sd   upbd   lwbd upbd_q lwbd_q upbd_b lwbd_b
 0.108  0.153  0.148  0.342 -0.037  0.359  0.009  0.401 -0.039

$$indirect.effect
         y1.all y1.tvhours y1.cmpthours y1.cellhours y1.exercises
```

```
est      0.063        0.002          0.008        0.001          0.002
mean     0.132        0.006          0.023        0.004          0.004
sd       0.070        0.011          0.036        0.016          0.029
upbd     0.221        0.020          0.069        0.024          0.041
lwbd     0.043       -0.008         -0.023       -0.016         -0.033
upbd_q   0.208        0.018          0.053        0.021          0.031
lwbd_q   0.064       -0.004         -0.023       -0.012         -0.026
upbd_b   0.212        0.035          0.057        0.030          0.054
lwbd_b   0.011       -0.016         -0.039       -0.026         -0.039
         y1.sweat  y1.sports   y1.j1
est        0.008      0.043    0.024
mean       0.005      0.102    0.032
sd         0.015      0.063    0.046
upbd       0.024      0.183    0.091
lwbd      -0.015      0.021   -0.028
upbd_q     0.021      0.190    0.079
lwbd_q    -0.007      0.043   -0.008
upbd_b     0.036      0.173    0.132
lwbd_b    -0.026      0.017   -0.013
```

If the argument *RE* is set as "TRUE", *summary* reports the summaries of the relative effects, calculated as the "(in)direct effect/total effect". The confidence level is set to be $1 - \alpha$ where α is set by the argument *alpha*, which is 0.05 by default.

```
> summary(bootmed.b.b.2,RE=T,alpha=0.2,quant=T)
MMA Analysis: Estimated Mediation Effects Using MART
The relative effects:
For Predictor/Moderator at predF
$$direct.effect
   est    mean      sd    upbd    lwbd  upbd_q  lwbd_q  upbd_b  lwbd_b
 0.633   0.410   0.374   0.890  -0.069   0.765   0.082   0.878  -0.500

$$indirect.effect
         y1.all  y1.tvhours  y1.cmpthours  y1.cellhours  y1.exercises
est       0.367       0.011         0.045         0.005         0.011
mean      0.590       0.032         0.068         0.009         0.003
sd        0.374       0.056         0.233         0.084         0.188
upbd      1.069       0.105         0.366         0.117         0.244
lwbd      0.110      -0.040        -0.231        -0.098        -0.239
upbd_q    0.918       0.113         0.252         0.089         0.154
lwbd_q    0.235      -0.022        -0.070        -0.082        -0.095
upbd_b    1.500       0.181         0.659         0.193         0.364
lwbd_b    0.122      -0.080        -1.002        -0.211        -0.837
         y1.sweat  y1.sports   y1.j1
est         0.049      0.250    0.144
mean        0.015      0.526    0.128
sd          0.088      0.526    0.213
upbd        0.128      1.200    0.401
```

```
lwbd      -0.098     -0.148 -0.145
upbd_q     0.088      1.095  0.311
lwbd_q    -0.022      0.140 -0.083
upbd_b     0.250      2.498  0.969
lwbd_b    -0.132      0.097 -0.199
```

Using the quantile confidence interval, *summary* shows that at the significance level 0.2, only "sports" has a significant indirect effect on the sexual difference in overweight, which explains about 52.6% (14.0%, 109.5%) of the sexual difference. After accounting for all the factors, the direct effect is still significant different from 0: about 41% (8.2%, 76.5%) of the sexual difference cannot be explained. When we set *plot=T* in the *summary* function, by default, the function plots the estimated third-variable effects and their confidence intervals. By default, the confidence intervals from the normal approximation, *(lwbd, upbd)*, are drawn. If we set *quant=T*, the quantile confidence intervals, *(lwbd_q, upbd_q)* are used. When *ball.use=T*, the ball confidence intervals, *(lwbd_b, upbd_b)*, are drawn. Figure 5.1 is a product from the summary function above.

The *plot* function helps illustrate the indirect effects of each third-variable by showing the relationship between the third-variable and the outcome after accounting for other variables. The plot also shows how each exposure variable relates to the third-variable. If the exposure variable is continuous, the fitted relationship between the exposure and the third-variable is plotted. Otherwise, the distributions of the third-variable at different levels of the exposure are graphed. The third-variable to be plotted is specified by the argument *vari*, which can be the column number or the name of the third-variable in x. We can also change the range of the third-variable by setting the argument *xlim*. The following codes illustrate how the continuous third-variable *exercise* intermediates the sexual difference in overweight. The plotting results are Figures 5.2 and 5.3. The fitted relationship between overweight and exercise is not linear since a MART was used to model the relationship.

```
plot(bootmed.b.b.2,vari="exercises",xlim=c(1,50))
```

All codes are provided in the book website. In addition, we have written a vignette for the package, which is obtainable through the command `browseVignettes("mma")`.

5.2 SAS Macros

For SAS users, we developed SAS macros so that the multiple third-variable analysis can be performed utilizing the R package, *mma* in the SAS environment. As with the *mma* R package, the TVE analysis can be performed by steps or in one step.

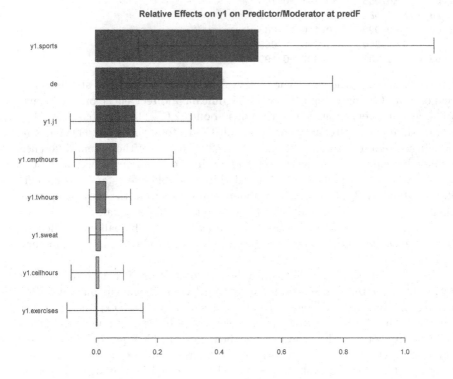

FIGURE 5.1
The plot from the function *summary(bootmed.b.b.2,RE=T,alpha=0.2, quant=T)*.

This section is derived in part from an article published in the Journal of Open Research Software [20], available online: https://openresearchsoftware. metajnl.com/articles/10.5334/jors.277/.

5.2.1 Running R in SAS

In order to run R packages in SAS, SAS version 9.22 or above is required. To communicate with R, an RLANG option needs to be set when starts SAS. An RLANG is set through modifying the *sasv9.cfg* file, which is typically located at *C:\Program Files\SASHome\x86\SASFoundation\9.4*. The path can be different according to where the SAS is installed and what is the version of SAS. To edit the file, a user has to run the operating system (e.g. Windows) as an administrator. Once the file is located, three lines are to be edited as follows.

```
-RLANG
-config "C:\Program Files\SASHome\x86\SASFoundation\9.4\
```

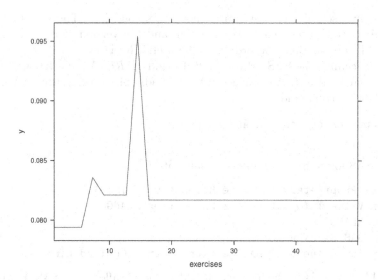

FIGURE 5.2
The relationship between overweight and sports.

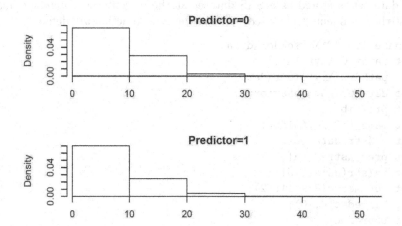

FIGURE 5.3
The exercise level for boys (predictor=0) and girls (predictor=1) separately.

```
         nls\en\sasv9.cfg"
-SET R_HOME "C:\Program Files\R\R-3.5.3"
```

Usually users do not have to change the second line. The first line and third line need to be added. The location and the version of R in the third line need to be modified according to the users' R setting.

Next, change the SAS "Properties" by adding *RLANG* to the end of the target command line. And then run SAS as administrator. Within SAS, run the following command:

```
proc options option = RLANG value;
run;
```

In the SAS log, a message appears that states:

```
SAS (r) Proprietary Software Release 9.4  TS1M1
Option Value Information For SAS Option RLANG
Value: RLANG
Scope: SAS Session
How option value set: SAS Session Startup Command Line
```

If *RLANG* appears in the message, R codes can be called from SAS. For more information about calling R within SAS, readers are referred to [35].

5.2.2   Macros to Call the *data.org* Function

A couple of macros are prepared to help SAS users call *data.org* from SAS. We still use the "weight_behavior" data set as the example. In the example, the data set is stored as a SAS dataset at the directory *C:\myfolder\data*. First the arguments to be used with the data.org function are defined:

```
libname lib "C:\myfolder\data";
%let path= C:\myfolder;
%let pathd=C:\myfolder\data;
%let data=weight_behavior;
%let pre=lib.;
%let path_r=C:/myfolder;
%let x=%str(data[,c(2,4:14)]);
%let pred=%str(data[,3]);
%let y=%str(data[,15]);
%let contmed=c(7:9,11:12);
%let binmed=c(6,10);
%let binref=c(1,1);
%let catmed=5;
%let catref=1;
%let predref=M;
%let alpha=0.4;
%let alpha2=0.4;
```

```
Identified as mediators:
[1] "exercises" "sweat"      "sports"
Selected as covariates:
[1] "age"        "race"      "numpeople" "car"        "cellhours"
Tests:
               P-Value 1.y P-Value 2.pred
age               0.836               NA
race              0.850               NA
numpeople         0.561               NA
car               0.052               NA
gotosch           0.710               NA
snack             0.724               NA
tvhours           0.830               NA
cmpthours         0.826               NA
cellhours         0.067            0.688
sports *          0.000            0.003
exercises *       0.176            0.203
sweat *           0.181            0.046
pred              0.178               NA
----
*:mediator,-:joint mediator
P-Value 1:Type-3 tests in the full model (data.org) or estimated coefficients (data.org.big)
when testtype=1, univariate relationship test with the outcome when testtype=2
P-Value 2:Tests of relationship with the Predictor
```

FIGURE 5.4
SAS log screen for the data.org function.

We include in the book website a template and example for setting up all arguments that are used for the third-variable analysis. The macro is called "setup_mma_macro_wb".

Once the arguments are defined, the macro *Proc_R_dataorg* on website/appendices helps read in all arguments and execute the functions in the *R_submit_dataorg.sas* in the R setting. Both macros should be stored in the directory defined by *path*. The macro is executed by the following command:

```
%include "&path\Proc_R_dataorg.sas";
```

Lastly, run the *Proc_R_dataorg* macro:

```
%Proc_R_dataorg(&pre, &path_r, &data, &mediator, &contmed,
&binmed, &binref, &catmed, &catref, &predref, &refy, &alpha,
&alpha2, &x, &pred, &y);
```

The results from the *summary* function of the output of *data.org* is printed on the SAS log screen as in Figure 5.4 and the output of *data.org* is stored in the directory defined by *path_r*.

5.2.3 Macros to Call the *med* Function

The third-variable analysis with the function *med* is based on the outputs from the function *data.org*. As the first step, arguments to be used in the function are defined:

```
%let rdata=data.bin;
%let n=10;
%let nonlinear=FALSE;
```

The template to input values for arguments of *med* is also provided in the file *setup_mma_macro.sas* on website/appendices. And then the macro *Proc_R_med.sas* helps read in all arguments and the macro *R_submit_med* defines the functions to be executed in R. The files *Proc_R_med.sas* and *R_submit_med.sas* are provided on website/appendices and should be stored in the *path* defined above.

```
%include "&path\Proc_R_med.sas";
```

Lastly, run the *Proc_R_med* macro:

```
%Proc_R_med(&pre,&path_r,&rdata,&margin,&D,&distn,&refy,
&n,&nu,&nonlinear,&df1,&type)
```

The outputs are saved as an R dataset in the path defined by *path_r*. The analysis results are printed in the SAS log file.

5.2.4   Macros to Call the *boot.med* Function

Similarly to the execution of functions *data.org* and *med*, arguments to be used in the function *boot.med* are defined first. For example, to use the results from *data.org* above, and run bootstrap samples for 40 times using generalized linear models, we define the arguments in the following way:

```
%let rdata=data.bin;
%let n=10;
%let n2=40;
%let nonlinear=FALSE;
```

Again, the template to assign arguments are provided in the file *setup_mma_macro.sas* on website/appendices. The macro *Proc_R_bootmed.sas* helps read in all arguments and the macro *R_submit_bootmed* defines the functions to be executed in R. The files *Proc_R_bootmed.sas* and *R_submit_bootmed.sas* are provided on website/appendices and should be stored in the *path* defined above.

```
%include "&path\Proc_R_bootmed.sas";
```

Lastly, run the *Proc_R_bootmed* macro:

```
%Proc_R_bootmed(&pre, &path_r, &rdata, &margin, &D,
&distn, &refy, &n, &n2, &nu, &nonlinear, &df1, &type);
```

The outputs are saved as an R dataset in the path defined by *path_r*. The summary results of the output, an *mma* object, are printed in the SAS log file. A plot similar to 5.1 is saved as *data_bin_plot.png* in the path defined by *path_r*, if *plot = F* is not specified in the arguments.

5.2.5 Macros to Call the *plot* Function

To plot the relationship of how a third-variable intermediates the relationship between the exposure and the outcome, a user needs to first specify the third-variable of interests by:

```
%let vari=exercises;
```

The macros *Proc_R_plot.sas* helps read in all arguments and the macro *R_submit_plot* defines the functions to be executed in R. The files *Proc_R_plot.sas* and *R_submit_plot.sas* are provided at the appendices and on the course website. It should be stored in the *path*. Run the following commands:

```
%include "&path\Proc_R_plot.sas";
%Proc_R_plot(&pre, &path_r, &vari, &alpha, &quantile,
 &xlim);
```

As results, the plots similar to Figures 5.2 and 5.3 are saved in the path defined by *path_r*, as the file named "data.bin.plot2.png". The file name can be changed in the *R_submit_plot* macro.

5.3 Examples and Simulations on General Third-Variable Effect Analysis

In this section, we provide examples on using the general third-variable analysis method in practice. And then we use simulation studies to check the sensitivity and power of identifying important third-variables using the method. The results of the examples have been published in [79] and [86].

5.3.1 Pattern of Care Study

A series of studies referred to as Pattern of Care studies (PoC) were conducted by the National Program of Cancer Registries of the Centers of Disease Control and Prevention. The studies were initiated in response to findings reported by the Institute of Medicine that cancer patients do not consistently receive care known to be effective for their conditions. The PoC-Breast and Prostate (BP) study is one of the studies that collected complete health care information and tumor characteristic information for female breast cancer patients. The study collected patient care and follow-up information from breast cancer patients and subsequently linked with data from a number of sources, including 1) census data for social-environmental information, 2) hospital and physician files for measures of health systems and 3) hospital, provider, and Medicare files for obtaining additional information on comorbidity and adjuvant treatment.

The Louisiana Tumor Registry participated in the PoC-BP study, which re-abstract medical records of 1453 Louisiana non-Hispanic white and black women diagnosed with invasive breast cancer in 2004 [79]. All patients were followed up for five years or until death, whichever is shorter. We found that the odds of dying of breast cancer within three years for black women was significantly higher than that for white patients ($OR = 2.03, 95\%CI$: $(1.468, 2.809)$). To explain the racial disparity in breast cancer mortality rate or survival, we apply the third-variable analysis method on the PoC-BP study at Louisiana to separate effects from different risk factors. In Sections 5.3.2, the outcome is the three year mortality. In Section 5.3.3, the outcome is the censored survival time. The variables used in this study and their values, format, and data sources are listed in Table 5.2.

Tables 5.3 and 5.4 show some summary statistics of variables at different race groups among the 1453 patients. The p-values are the test results on the null hypothesis that the distributions of each variable are equivalent between the White and African American populations. The results are obtained through the output of the function *data.org* in the R package *mma*.

5.3.2   To Explore the Racial Disparity in Breast Cancer Mortality Rate

In this section, we use the PoC-BP data from the Louisiana Tumor Registry to explore the racial disparity in the three-year mortality rate among Louisiana breast cancer patients. In this example, the explanatory variable is the race of patient (0 for white and 1 for black). The binary response variable is patient's vital status at the end of the third year of diagnosis (death from breast cancer or not). First, we use the data.org function to identify third-variables and covariates. In the following code, x is a data frame that includes all variables listed in Tables 5.3 and 5.4. y is the outcome vector, where $y = 1$ if the patient died within three-year of breast cancer diagnosis and $y = 0$ otherwise. The *pred* is the exposure variable indicating the race (White or Black) of the patient. All variables in x are tested so that they are either removed or included as third-variables or covariates. The significance levels are set at 0.05.

```
data.poc.mortality<-data.org(x,y,pred=pred,mediator=
                1:ncol(x), alpha=0.05,alpha2=0.05)
```

Since the outcome, three-year mortality rate is binary, we use both the logistic regression and MART to explore the third-variable effect. We use the bootstrap method with 1000 re-samples.

```
bootmed.mortality.1<-boot.med(data=data.poc.mortality,
                n=20, n2=1000)
bootmed.mortality.2<-boot.med(data=data.poc.mortality,
            n=20, n2=1000,nu=0.05,nonlinear=TRUE)
summary(bootmed.mortality.1)
```

TABLE 5.2
List of variables and data sources.

Group	Variable Name (values/formats)	Data Source
Outcome	date of last contact, vital status, cause of death (right-censored data)	Cancer registry, PoC study
Patient Information	insurance status (private, public, no)	Hospital discharge sheets, Medical records, Cancer registry
	race (black, white)	
	age at diagnosis (continuous, *years*)	
	marital status (single, married, separated/divorced/widowed)	
	BMI (continuous, kg/m^2)	
	comorbidity (no/mild, moderate, severe)	
Tumor Characteristics	tumor grade (categorical)	Cancer registry, PoC
	lateral (left, right)	
	tumor size (continuous, *mm*)	
	lymph nodes involvement (yes, no)	
	extension (from the primary and distant metastases: yes, no)	
	cancer type (four types by ER/PR, her2 (+,-))	
	stage (SEER stage 2000)	
Environment Variable	poverty ($\geq$ 20% vs < 20% households below poverty)	Census
	education ($\geq$ 25% vs < 25% adults with less than high school)	
	residence area (rural, mixed, urban)	
	employment ($\geq$ 66% vs < 66% adults unemployed	
Treatment Information	surgery (no, lumpectomy, mastectomy)	PoC
	radiation (no, yes)	
	chemo-therapy (no, yes)	
	hormonal (no, yes)	
Health Care Facility	bed size (continuous)	American Hospital Association (AHA), PoC Study
	hospital ownership (public, private)	
	teaching hospital (yes, no)	
	COC status (yes, no)	

TABLE 5.3
Potential categorical mediators/confounders and covariates.

Potential Mediators/confounders	Black	White	P-value
insurance: no insurance	33.00%	13.06%	< .0001
public insurance	29.95%	9.28%	
private insurance	62.32%	87.51%	
marital status: single	33.00%	13.06%	< .0001
married	38.30%	59.18%	
separated or divorced	9.62%	7.41%	
widowed	19.07%	20.35%	
comorbidity: no/mild	36.08%	49.00%	< .001
moderate	48.93%	39.55%	
severe	14.99%	9.92%	
tumor grade: I	9.23%	17.20%	< .001
II	37.64%	47.63%	
III	49.82%	32.61%	
IV	3.32%	2.57%	
lateral: left	49.76%	52.52%	.3173
lymph nodes involvement: yes	55.61%	62.09%	.0156
extension:yes	89.43%	93.40%	.008
cancer subtype: ErPr-, Her2-	23.85%	12.25%	< .001
ErPr+, Her2-	39.58%	50.25%	
ErPr-, Her2+	12.19%	9.72%	
ErPr+, Her2+	24.38%	27.78%	
stage: 1. Localized	50.41%	58.50%	< .001
2. Regional by direct extension only	1.78%	1.85%	
3. Ipsilateral regional lymph node(s) only	31.60%	27.86%	
4. Regional by both 2 and 3	5.19%	5.20%	
7. Distant sites	11.02%	6.59%	
poverty: ≥ 20%	71.13%	26.80%	< .001
education: ≥ 25%	71.77%	39.06%	< .001
residence area: rural	62.58%	46.16%	< .001
mixed	6.29%	13.52%	
urban	31.13%	40.32%	
employment: ≥ 66%	81.29%	53.61%	< .001
surgery: no	9.24%	4.05%	< .001
lumpectomy	45.06%	46.99%	
mastectomy	45.71%	48.96%	
radiation: no	56.04%	53.38%	.335
chemo-therapy: no	54.81%	68.83%	< .001
hormonal: no	60.86%	48.89%	< .001
hospital ownership: private	77.2%	74.42%	.259
teaching hospital: yes	53.54%	39.27%	< .001
COC status: yes	64.35%	64.66%	.948

Note: P-value is the result of testing the association between race and the row
variables.

TABLE 5.4
Potential continuous mediators/confounders and covariates.

Potential Mediators	Black Mean (sd)	White Mean (sd)	P-value
age at diagnosis	57.95(14.08)	61.47(13.65)	< .0001
BMI	32.52(8.06)	29.24(7.08)	< .0001
tumor size	30.15(26.48)	22.73(18.68)	< .0001
bed size	3.40(1.43)	3.31(1.38)	.229

Note: P-value shows the result of testing the association between race and the row variables.

```
summary(bootmed.mortality.2)
```

The results from both the linear and nonlinear models are summarized in Table 5.5. Note that the IEs were measured in terms of log odds of death in logistic model, but in terms of probability of death in MART. To compare results, we define the relative indirect effect: $RE = \frac{IE}{TE}$. Table 5.5 presents the indirect effect, relative indirect effect estimates, and their 95% confidence intervals from the bootstrap method.

There are some differences in the results from logistic regression and from MART in that:

1. the order of relative effects are slightly different;

2. comorbidity is a significant third-variable by MART but not by logistic regression;

3. the confidence intervals were narrower using MART.

We recommend using MART in this case, since it is more sensitive in finding significant third-variables especially if the assumed relationship in logistic regression is inappropriate. From the results of MART, the stage explains 29% of racial disparities in the three-year mortality rate among breast cancer patients. Compared with localized breast cancer, patients diagnosed with regional or distant cancer are significantly more likely to die from breast cancer within three years. By Table 5.3, the probability of diagnosed with more advanced stages is significantly higher $(p-value < 0.0001)$ among black patients than among white patients. Also, the larger tumor size and/or worse grade of breast tumor were related with higher risk of breast cancer death, which were also more prevalent among black women than among white women. The relative indirect effects from tumor size and grade were 21% and 10%, respectively.

Compared with patients with negative ER/PR receptors, patients with positive ER/PR receptors develop less aggressive breast cancer and have better three-year survival rates. Black women are 54.7% less likely to be diagnosed with positive ER/PR receptors than whites. Compared with patients with lumpectomy surgery or hormonal therapy after surgery, those without

TABLE 5.5
Indirect effects (IE) and relative effects (RE).

Third-variable	Logistic Regression		Nonparametric Method	
	IE(95% CI)	RE(95% CI)	IE(95% CI)	RE(95% CI)
Stage	0.276(0.127,0.488)	28.14(12.2,76.7)	0.023(0.008,0.038)	28.78(9.6,48.0)
Insurance	0.275(0.074,0.430)	28.05(5.8,70.8)	0.006(0.004,0.008)	7.84(5.6,10.1)
ER/PR	0.181(0.068,0.332)	18.46(5.5,56.0)	0.015(0.009,0.020)	18.53(11.8,25.3)
Grade	0.158(0.027,0.379)	16.09(2.4,47.6)	0.008(0.005,0.011)	10.34(6.23,14.5)
Surgery	0.145(0.067,0.333)	14.75(6.3,50.4)	0.016(0.008,0.023)	19.87(10.4,29.3)
Tumor Size	0.135(0.025,0.417)	13.77(2.2,56.1)	0.016(0.011,0.022)	20.88(14.3,27.5)
Hormonal Therapy	0.114(0.02,0.253)	11.57(1.6,40.2)	0.008(0.004,0.011)	9.75(5.6,13.9)
Age	-0.113(-0.301,-0.034)	-11.5(-43.2,-3.1)	-0.005(-0.011,0.001)	-6.04(-13.6,1.6)
Marital Status	0.001(-0.001,0.004)	1.65(-1.27,4.57)	0.001(-0.001,0.004)	1.65(-1.3,4.6)
Comorbidity	-0.002(-0.150,0.109)	-0.17(-20.5,13.6)	0.002(0.001,0.003)	3.06(1.9,4.2)

[1] After considering all indirect effects through third-variables, direct effect of race on mortality was -0.115 with 95% CI(-0.713,0.526)in logisitic model and was -0.014(-0.034,0.007).
[2] 95% confidence interval is 0.025 and 0.975 percentiles of the distribution of statistics obtained by bootstrap with 1000 repetitions.

surgery or hormonal therapy have worse three-year survival. Black women are less likely to undergo hormonal therapy or lumpectomy surgery. As results, the relative indirect effects for hormonal receptor, surgery, and hormonal therapy are 10%, 20% and 10%, respectively.

Black breast cancer patients are more likely to be diagnosed at younger age at which the patients have longer survival time. This fact indicates that age at diagnosis is a suppression factor for racial disparities in mortality.

Insurance has a significant indirect effect (8%) on racial disparities in the risk of death from breast cancer. Compared with patients with private insurance, those having no insurance or having Medicaid, Medicare or other public insurance have higher three-year mortality, which might relate to the more restricted accessibility to necessary breast cancer treatment. A higher proportion of blacks than whites in this study have Medicaid, Medicare, public or no insurance.

The estimated average direct effect of race on mortality after considering all third-variables is $-.014$ with 95% confidence interval $(-.034, .007)$, which was statistically insignificantly different from 0. This suggests that racial disparities of breast cancer survival are satisfactorily explained by all third-variables in the model.

5.3.3   To Explore the Racial Disparity in Breast Cancer Survival Rate

We then use the time of survival as the outcome. The outcome is the right-censored number of days after diagnosis of breast cancer. In the following code, *brdata* is the data set with all variables collected for the PoC study, in which *fudays* is the number of days for follow up and *cdeath* is the event (death) status. *x* and *pred* are the same as in Section 5.3.2. In addition, variables that are theoretically considered as critical in explaining the racial disparity are forced into the analysis as third-variables without statistical tests. We forced in three groups of related variables as third-variables: 1) the health care facility variables: *bed size, hospital ownership, teaching status,* and *COC status,* 2) census tract social economic factors: *poverty, education,* and *employment* and 3) cancer stage related variables: *stage, lymph nodes involvement, extension,* and *tumor size.*

```
y=Surv(brdata1$fudays,brdata1$cdeath)
data.surv.breast<-data.org(x,y,pred,mediator=(1:ncol(x)),
 alpha=0.05,alpha2=0.05,jointm = list(n=3,j1=c("bed",
 "coc","own","tch"),j2=c("emp","edu","poverty"),
 j3=c("SS2000", "size", "nodes", "ext")))
```

For the time-to-event outcome, we use both the COX hazard model and MART to explore the third-variable effect. We use the bootstrap method with 1000 re-samples.

```
bootmed.survival.1<-boot.med(data=data.poc.sruvival,n=20,
                        n2=1000)
bootmed.survival.2<-boot.med(data=data.poc.survival,n=20,
                        n2=1000,nu=0.05,nonlinear=TRUE)
summary(bootmed.survival.1)
summary(bootmed.survival.2)
```

Table 5.6 shows the estimated direct and indirect effects in explaining the racial disparity in breast cancer survival for Louisiana patients who were diagnosed with breast cancer in 2004 using the linear (Cox proportional hazard model and generalized linear model) and nonlinear (MART survival model and smoothing splines) models, respectively. For both models, the racial disparity in breast cancer survival is completely explained by all included variables because for both linear and nonlinear methods, the estimated direct effect is not significantly different from 0. Figure 5.5 is the results from the summary of the nonlinear model.

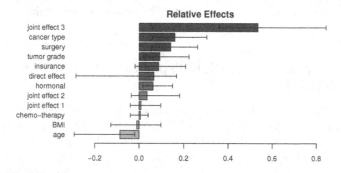

FIGURE 5.5
Relative effects from non-linear models.

Compared with whites, non-Hispanic blacks have an average higher hazard rate by both linear ($TE = 0.629, 95\%$ CI $(0.135, 2.815)$) and non-linear ($TE = 0.515, 95\%$ CI $(0.392, 0.971)$) models. From Table 5.6, we see that the confidence intervals for estimated (in)direct effects are much wider using linear models than using nonlinear models. This is due to the inclusion of many highly correlated variables in predicting the hazard rate. The multi-collinearity results in highly variant estimates from linear models. MART accounts for potential nonlinear relationships in hazard prediction and is much more robust to multi-collinearities. Again, we explain the results using the nonlinear models. On average, the instantaneous death at any time is $67.36\%(e^{0.034} - 1)$ higher among blacks than among whites. The third-variables in Table 5.6 are ordered according to the absolute value of their estimated relative effects from MART.

TABLE 5.6
Summary of mediation/confounding effect estimations for breast cancer survival.

Third-Variables	Linear Models		Nonlinear Models	
	IE (95% CI)	RE (%)	IE (95% CI)	RE (%)
joint effect 3	0.276 (0.109,0.528)	43.9 (7.3, 162.7)	0.275 (0.185, 0.556)	53.4 (36.0, 84.3)
molecular subtype	0.264 (0.153, 0.690)	41.9 (14.2, 161.9)	0.082 (0.038, 0.179)	15.9 (6.2, 30.3)
surgery	0.016 (−0.063, 0.088)	2.6 (−14.2, 22.4)	0.073 (0.031, 0.179)	14.1 (5.5, 26.2)
tumor grade	0.135 (−0.012, 2.203)	21.5 (−2.9, 107.5)	0.048 (0.011, 0.034)	9.2 (1.8, 22.3)
age at diagnosis	−0.166 (−0.271,−0.043)	−26.4 (−87.6, −2.6)	−0.045 (−0.171, −0.014)	−8.8 (−29.6, −1.9)
insurance	0.151 (−0.001, 0.325)	23.9 (−2.3, 80.6)	0.045 (−0.01, 0.136)	8.7 (−1.9, 20.8)
hormonal	0.131 (−0.012, 0.335)	20.9 (−2.5, 120.3)	0.033 (0.009, 0.095)	6.3 (1.6, 14.9)
joint effect 2	0.075 (−0.228, 0.379)	11.9 (−36.7, 162.7)	0.019 (−0.020, 0.124)	3.6 (−3.6, 18.1)
joint effect 1	−0.034 (−0.165, 0.118)	−5.5 (−43.0, 14.2)	0.004 (−0.025, 0.074)	0.9 (−4.1, 9.7)
chemo-therapy	−0.087 (−0.266, 0.015)	−13.8 (−66.2, 2.8)	0.003 (−0.024, 0.025)	−0.6 (−4.1, 4.1)
bmi	−0.084 (−0.298, 0.266)	−13.4 (−92.5, 23.4)	−0.006 (−0.069, 0.061)	−1.1 (−12.8, 9.8)
direct effect	−0.058 (−0.732, 0.563)	−9.3 (−355.0, 66.1)	0.034 (−0.163, 0.133)	6.6 (−28.8, 16.7)
total effect	0.629 (0.135, 2.815)		0.515 (0.392, 0.971)	

Note: the mediators/confounders are ordered according to the absolute value of the estimated relative effect (RE) with nonlinear models. The variables with significant indirect effects (IE) are boldfaced. *joint effect 1* is the joint effect from the health care facility variables, *joint effect 2* from the census tract social economics factors, and *joint effect 3* from the cancer stage related variables.

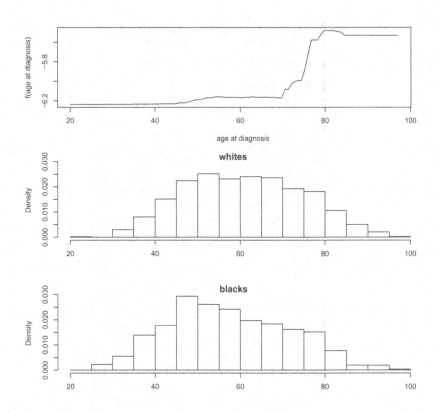

FIGURE 5.6
Indirect effect of *age* on *breast cancer hazard.*

"direct effect" is the racial disparity that cannot be explained by all the third-variables included in the model. In Table 5.6 we find that the estimated 95% confidence interval of direct effect is $(-0.163, 0.133)$, which includes 0. The cancer stage-related variables (*stage, tumor size, lymph nodes involvement, extension,* and *tumor size*), *joint effect 3,* explain about 53.4% of the racial disparity in survival. Other variables such as *molecular subtype* (15.9%), *surgery* (14.1%), *tumor grade* (9.2%), and *hormonal therapy* (6.3%) also significantly explain the racial difference. An interesting variable is *age at diagnosis,* which has a negative indirect effect (opposite to the total effect) and a significant negative relative effect (-8.8%). To gain some knowledge from the relationship, we plot the effect directions using the following codes:

```
plot(bootmed.survival.2,vari=''age'')
```

The result is Figure 5.6. We find that the hazard rate increased with age at diagnosis (upper panel of Figure 5.6), and blacks tended to be diagnosed at younger ages than whites. Therefore, accounting for age would enlarge the gap in hazard rates between blacks and whites. The middle and lower panels of Figure 5.6 show the age distributions among whites and blacks, respectively.

Similarly, Figure 5.7 describes how *tumor size* are drawn using the *plot* function and can explain the racial disparity in survival. The top plot shows the relationship between *tumor size* and the hazard rate. The line was fitted using the MART. We can see that the hazard increases as the tumor size increases. The lower two plots show the *tumor size* distributions among white (middle plot) and black (lower plot) populations, respectively. We found that whites are diagnosed with cancer at a relatively smaller tumor size than are blacks. More than half (53.4%) of the racial disparity is explained by the cancer stage-related variables (joint effect 3).

Figure 5.8 shows how the categorical variable *molecular subtype* helps explain the racial disparity in cancer survival. On average, ER or PR negative has an average higher hazard rate than ER/PR positive patients, and there are more black patients are diagnosed with ER/PR negative (36.04% in blacks vs. 21.97% in whites). Therefore, the molecular subtype explains 15.9% of the racial disparity. We did not find significant joint effect from the census tract level social economics factors. Overall, diagnosed at an early age, at early stages, with ER/PR positive, had surgery, with low tumor grade, and had hormonal therapy are related with high survival rate.

5.3.4 Simulation Study

To evaluate the sensitivity and specificity of the proposed method, we conduct some simulations to check the bias, type I error, and power in estimating third-variable effects. We check the method comprehensively with all types of exposures, third-variables and response variables, and with many types of relationships. In this section, we show a special case where all variables were binary. For more scenarios, readers are referred to [19].

In this case, there are two potential third-variables, which were independent or correlated with each other. Data are generated from the following true models:

$$logit(Pr(M_1 = 1)) = a_1 X$$
$$logit(Pr(M_2 = 1)) = a_2 X$$
$$logit(Pr(Y = 1)) = cX + b_1 M_1 + b_2 M_2$$

The correlation between M_1 and M_2 was controlled by the odds ratio $OR = \frac{Pr(M_1=1,M_2=1) \cdot Pr(M_1=0,M_2=0)}{Pr(M_1=0,M_2=1) \cdot Pr(M_1=1,M_2=0)}$ [56]. Four different sample sizes of the simulation (200, 500, 1000 and 1500) are applied. Parameters a_1 and a_2 are chosen from 0, 0.518 or 2.150, which indicates different proportions of variances in M that is explained by X. Without loss of generality, we make $a_1 \leqslant a_2$. Thus there are six different combinations of (a_1, a_2). Parameters c and OR took values from the sets $(0, 0.518, 2.150)$ and $(0.2, 1, 5)$, respectively. The values of

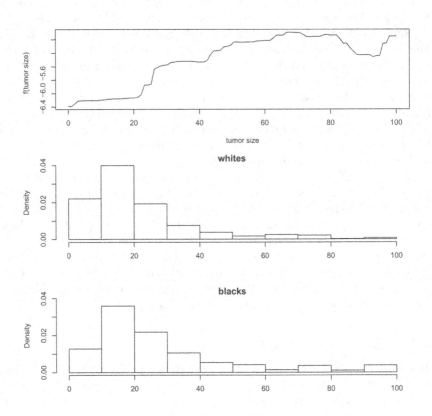

FIGURE 5.7
Indirect effect of *tumor size* on *breast cancer hazard.*

b_i ($i = 1, 2$) depend on a_i ($i = 1, 2$). When $a_i = 0$, $a_i = 0.518$ or $a_i = 2.150$, b_i takes values from the sets $(0, 0.518, 2.150)$, $(0, 0.522, 2.168)$ or $(0, 0.564, 2.341)$, respectively. Larger a_j and b_j indicates greater indirect effect from M_j. When either of them is zero, there is no mediation effect from the corresponding variable. A total of $6 \times 3^4 = 486$ parameter combinations are used for each of the four sample sizes, yielding 1944 simulation scenarios. We present the simulation results for $a_1 = 0.518, a_2 = 0.518$ and sample sizes 500 and 1000 here.

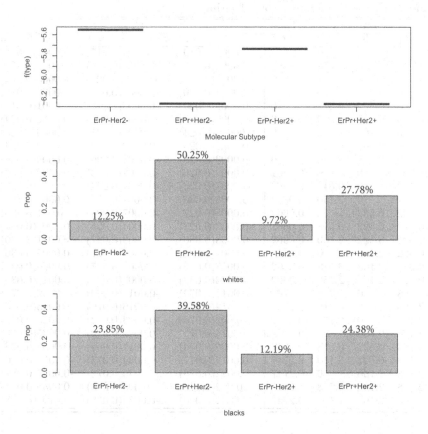

FIGURE 5.8

Indirect effect of *cancer type* on *breast cancer hazard.*

5.3.4.1 Empirical Bias

For each simulation scenario, 500 replications are conducted. We estimate the IEs of M_1 and M_2 from these 500 replicates. The empirical bias is the difference between the averaged IE and true IE. The simulation results are summarized in Table 5.7. For all the scenarios, we find no empirical bias that is significantly different from 0.

5.3.4.2 Type I Error Rate and Power

For each replication described above, the variance of the estimated IE is estimated from both the Delta method and the bootstrap method ($B = 500$). We

TABLE 5.7
Empirical bias (standard error) of estimated AIE_{M_1}

Parameters			True	Bias		
b_1	b_2	c	AIE	$OR = 0.2$	$OR = 1$	$OR = 5$
0	0	0	0	0.0001(0.019)	0.0004(0.017)	0.0007(0.018)
0	0	0.518	0	-0.0003(0.019)	0.0003(0.017)	0.003(0.018)
0	0	2.150	0	-0.0004(0.023)	0.00004(0.021)	0.0008(0.023)
0	0.522	0	0	-0.0004(0.020)	-0.0001(0.017)	0.0002(0.019)
0	0.522	0.518	0	-0.0002(0.020)	-0.0002(0.018)	-0.0001(0.019)
0	0.522	2.150	0	-0.0009(0.024)	-0.0007(0.022)	-0.0002(0.024)
0	2.168	0	0	0.00004(0.024)	0.0003(0.021)	0.0009(0.023)
0	2.168	0.518	0	-0.0002(0.026)	-0.0002(0.023)	0(0.024)
0	2.168	2.150	0	-0.0003(0.028)	-0.0001(0.026)	-0.0005(0.028)
0.522	0	0	0.0662	-0.0002(0.026)	-0.0002(0.023)	-0.0002(0.025)
0.522	0	0.518	0.0662	-0.00007(0.026)	-0.0001(0.024)	-0.0002(0.026)
0.522	0	2.150	0.0662	-0.0005(0.030)	-0.0002(0.026)	0.0002(0.028)
0.522	0.522	0	0.0662	-0.0005(0.027)	-0.0007(0.024)	-0.001(0.026)
0.522	0.522	0.518	0.0662	-0.0003(0.027)	-0.0004(0.025)	-0.0006(0.027)
0.522	0.522	2.150	0.0662	-0.0008(0.030)	-0.0009(0.027)	-0.0011(0.030)
0.522	2.168	0	0.0662	-0.0001(0.032)	-0.0004(0.027)	-0.0005(0.030)
0.522	2.168	0.518	0.0662	0.0007(0.032)	-0.00003(0.027)	-0.0006(0.031)
0.522	2.168	2.150	0.0662	0.0002(0.033)	-0.0003(0.030)	-0.0016(0.035)
2.168	0	0	0.2747	-0.0011(0.074)	-0.0014(0.069)	-0.0022(0.074)
2.168	0	0.518	0.2747	-0.0012(0.074)	-0.0016(0.070)	-0.0023(0.075)
2.168	0	2.150	0.2747	-0.002(0.076)	-0.0023(0.072)	-0.0026(0.077)
2.168	0.522	0	0.2747	-0.0015(0.075)	-0.0023(0.070)	-0.0035(0.076)
2.168	0.522	0.518	0.2747	-0.0005(0.074)	-0.0017(0.069)	-0.0034(0.076)
2.168	0.522	2.150	0.2747	-0.0019(0.077)	-0.0034(0.073)	-0.0053(0.080)
2.168	2.168	0	0.2747	-0.0022(0.078)	-0.0029(0.072)	-0.0049(0.081)
2.168	2.168	0.518	0.2747	-0.002(0.078)	-0.003(0.073)	-0.0059(0.084)
2.168	2.168	2.150	0.2747	-0.0042(0.080)	-0.0055(0.078)	-0.0101(0.088)

test the hypothesis: $H_0 : IE = 0$ via the test statistics $\frac{IE}{SE(IE)}$, which is assumed to have a standard normal distribution. The type I error rate or power is the proportion of times rejecting the null hypothesis at 5% significance level. The simulation results are summarized in Table 5.8. If the true indirect effect is 0, the values listed in the table represent the type I error rates, otherwise the powers. Figure 5.9 presents the power curves obtained by the Delta and bootstrap methods ($OR = 0.2$) for M_1 at different b_1 and sample sizes when b_2 and c are fixed at 0.522 and 0.518, respectively. The Delta and bootstrap methods show similar patterns. Figure 5.9 suggests that for all sample sizes, the statistical power increases with b_1 and then reaches a plateau when b_1 hits a certain point. At fixed b_1, a larger sample size has greater power. For a given sample size, when b_2 and c are fixed, the statistical power increases as b_1 increases; when b_1 and b_2/c are fixed, the statistical power decreases as c/b_2 increases. At the same parameter configuration, indirect effect inference shows slightly greater power if two third-variables are generated independently.

TABLE 5.8
Type I error rates and power of estimated AIE_{M_1} by delta(bootstrap) method.

	Parameters		True	N = 500			N = 1000		
b_1	b_2	c	AIE	OR = 0.2	1	5	OR = 0.2	1	5
0	0	0	0	.008(.006)	.016(.012)	.008(.006)	.036(.028)	.028(.022)	.022(.018)
0	0	0.518	0	.012(.01)	.016(.012)	.012(.008)	.026(.024)	.02(.018)	.014(.01)
0	0	2.150	0	.006(.008)	.02(.01)	.01(.008)	.026(.022)	.026(.024)	.028(.022)
0	0.522	0	0	.004(.002)	.01(.01)	.014(.006)	.028(.026)	.026(.022)	.02(.012)
0	0.522	0.518	0	.01(.006)	.008(.008)	.014(.006)	.034(.03)	.028(.03)	.026(.022)
0	0.522	2.150	0	.012(.006)	.012(.006)	.01(.0014)	.034(.028)	.028(.028)	.03(.032)
0	2.168	0	0	.01(.006)	.01(.004)	.004(.004)	.024(.02)	.024(.022)	.026(.018)
0	2.168	0.518	0	.01(.006)	.01(.002)	.004(.002)	.036(.03)	.028(.022)	.034(.028)
0	2.168	2.150	0	.004(.004)	.012(.008)	.006(.006)	.026(.02)	.02(.024)	.03(.026)
0.522	0	0	.0662	.416(.356)	.444(.388)	.374(.31)	.87(.85)	.904(.876)	.888(.87)
0.522	0	0.518	.0662	.388(.334)	.432(.36)	.378(.33)	.848(.828)	.89(.75)	.868(.854)
0.522	0	2.150	.0662	.274(.214)	.29(.252)	.272(.22)	.718(.696)	.776(.874)	.712(.678)
0.522	0.522	0	.0662	.358(.304)	.39(.34)	.366(.302)	.846(.792)	.896(.846)	.876(.862)
0.522	0.522	0.518	.0662	.32(.256)	.368(.332)	.336(.284)	.818(.656)	.866(.72)	.85(.814)
0.522	0.522	2.150	.0662	.208(.17)	.264(.218)	.242(.19)	.69(.656)	.742(.72)	.686(.648)
0.522	2.168	0	.0662	.204(.164)	.26(.204)	.232(.194)	.668(.618)	.766(.722)	.718(.676)
0.522	2.168	0.518	.0662	.196(.168)	.252(.204)	.224(.176)	.6(.556)	.704(.656)	.634(.59)
0.522	2.168	2.150	.0662	.136(.106)	.174(.138)	.162(.106)	.48(.458)	.558(.512)	.516(.456)
2.168	0	0	.2747	.804(.794)	.818(.812)	.796(.788)	.972(.974)	.98(.974)	.978(.978)
2.168	0	0.518	.2747	.802(.794)	.816(.804)	.796(.78)	.97(.974)	.98(.974)	.978(.978)
2.168	0	2.150	.2747	.8(.772)	.812(.786)	.788(.748)	.97(.968)	.98(.97)	.978(.976)
2.168	0.522	0	.2747	.802(.794)	.816(.806)	.796(.778)	.97(.972)	.98(.974)	.978(.978)
2.168	0.522	0.518	.2747	.8(.786)	.814(.8)	.796(.764)	.97(.974)	.98(.972)	.978(.978)
2.168	.522	2.150	.2747	.796(.742)	.81(.738)	.782(.66)	.968(.968)	.978(.97)	.978(.974)
2.168	0.522	0	.2747	.798(.772)	.812(.78)	.786(.672)	.97(.97)	.98(.97)	.978(.976)
2.168	2.168	0.518	.2747	.792(.756)	.806(.736)	.766(.524)	.968(.968)	.978(.97)	.976(.972)
2.168	2.168	2.150	.2747	.774(.642)	.786(.496)	.744(.242)	.966(.966)	.974(.968)	.976(.93)

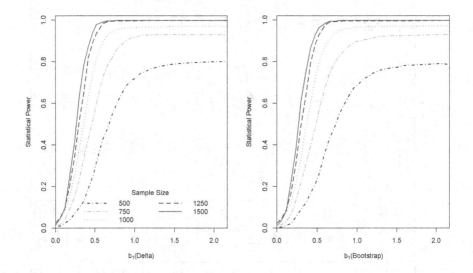

FIGURE 5.9
Statistical powers of estimating AIE_{M_1} from delta (left) and bootstrap (right) methods at different sample sizes ($b_2 = 0.522, c = 0.518$).

6

Assumptions for the General Third-Variable Analysis

There are three assumptions required for the general third-variable analysis as described in Section 4.2 and are restated here:

Assumption 1 *No-unmeasured-confounder for the exposure-outcome relationship.* This assumption can be expressed as $Y(x, \mathbf{m}) \perp\!\!\!\perp X|\mathbf{Z}$ for all levels of x and $\mathbf{m}$.

Assumption 2 *No-Unmeasured-Confounder in the X-M Relationship.* This assumption can be expressed as $X \perp\!\!\!\perp \mathbf{M}|\mathbf{Z}$ for all levels of x and $\mathbf{m}$.

Assumption 3 *No-unmeasured-confounder for the third variable-outcome relationship.* This assumption can be expressed as $Y(x, \mathbf{m}) \perp\!\!\!\perp \mathbf{M}|X, \mathbf{Z}$ for all levels of x and $\mathbf{m}$.

A fourth Assumption that is commonly used in the traditional third-variable analysis is not required in the general third-variable analysis proposed in Chapter 4. Since the assumption is typically required in many conventional third-variable analyses such as the methods based on linear models, we include it in the discussion.

Assumption 4 *Any third-variable M_i is not causally prior to other third-variables in $\mathbf{M}_{-i}$.*

In this chapter, we explain each of the assumptions and use simulations to study the situation when each of the assumptions is violated. We test the performance of estimations using the general third-variable analysis proposed in Chapter 4. This is the sensitivity analysis of assumptions. This chapter is derived in part from an article published in Communications in Statistics - Simulation and Computation [13] on April 8th, 2021, available online: https://doi.org/10.1080/03610918.2021.1908556.

To check the estimation results, we compare the accuracy and efficiency of third-variable effect estimates by biases, variances, the sensitivity and specificity rates. The sensitivity and specificity analysis is to check if significant third-variables can be correctly identified through the proposed method. For all calculations, the third-variable analyses are performed using the *mma* package. In Appendices, we include all codes for generating the series of simulations and accurately performing the third-variable analysis.

DOI: 10.1201/9780429346941-6

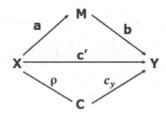

FIGURE 6.1
When the confounder C of the X-Y relationship is not observed.

6.1   Assumption 1: No-Unmeasured-Confounder for the Exposure-Outcome Relationship

This assumption requires that all third-variables be measured and put in the model appropriately. Figure 6.1 shows the relationship among four variables X, M, C and Y, where Y is the outcome and X is the exposure variable. A line with an arrow identifies a causal relationship between the two variables and a line without an arrow means that the two connected variables are related but no one is established as a causal variable to the other. In the figure, M is a mediator by definition and C is a confounder for the $X - Y$ relationship. By Assumption 1, all confounders for the exposure-outcome relationship should be included in analysis. If this is not the case, the estimation of third-variable effects can be biased.

In Figure 6.1, if C is not included in the analysis, we would like to check how that influences the estimations of direct and indirect effects. The data are simulated for both binary and continuous exposures. For the continuous exposure, we generate data in the following way:

$$\begin{pmatrix} X_i \\ C_i \end{pmatrix} \sim \ Bivariate\ Normal\left(\begin{pmatrix} 0 \\ 0 \end{pmatrix}, \begin{pmatrix} 1 & \rho \\ \rho & 1 \end{pmatrix}\right) \tag{6.1}$$

$$M_i = \alpha_1 + aX_i + \epsilon_{1i} \tag{6.2}$$

$$Y_i = \alpha_2 + bM_i + c'X_i + c_yC_i + \epsilon_{2i} \tag{6.3}$$

In the equations, $i = 1, \ldots, 50$, ϵ_{1i} and ϵ_{2i} are independent with standard normal distributions. α_1 and α_2 are intercepts for Equations 6.2 and 6.3 taking the values 0.2 and 0.1 separately. The correlation coefficient between X and C is ρ, choosing from the set $(-0.9, -0.5, -0.1, 0, 0.1, 0.5, 0.9)$ that includes various possible relationships from negative to positive. We are interested to see how different values of ρ can influence the estimation results. c' is the direct effect of X in this case and its value is chosen from the set $(0, 0.518, 1.402, 2.150)$, ranging from no effect to strong direct effect. c_y is selected from $(0, 0.259, 0.701, 1.705)$. In such setting, the indirect effect of M should be $a \times b$, where a is 0.773 and b is 0.701. To generate a binary exposure, we can simulate the exposure variable by setting it to be 0 if $X_i \leq 0$ and 1 if $X_i > 0$. We can use both the nonlinear and linear models to make inferences on the third-variable effects. In this chapter, we show results from the linear models with continuous exposures only. The codes for generating data and data analysis are included in the book website.

To check the performance of estimation, we calculate the biases and variances of estimates and the specificity or sensitivity of identifying a significant third-variable effect. Each simulation is repeated 100 times. The bias of an estimate is defined as the estimated value minus the true value of the effect. The bias reported is the average bias from the 100 simulations. A good estimate has an average bias close to 0. Sensitivity is defined as the probability of identifying a significant effect when the true effect is not 0. Specificity is the probability of finding an effect to be insignificant when the true effect is 0. A good estimate has both high sensitivity and specificity. A third-variable effect is identified as significant if the 95% confidence interval of the estimate does not include 0.

6.1.1   On the Direct Effect

Using the linear models, the bias of the estimated direct effect is shown in Figure 6.2. We found that no matter what is the true direct effect, c', the biases of the estimated direct effect are closely proportional to $\rho \times c_y$. This is because the confounder C was not included in the analysis and its effect is mixed into the estimated direct effect.

Next, we check the sensitivity and specificity of identifying significant direct effects. In Figure 6.3, when $c' = 0$, the upper left panel of the Figure shows the specificity of the method in identifying direct effects. We found that the specificity decreases when ρ is away from 0 and decreases when c_y is away from 0. This is due to that the bias of the estimated direct effect is $\rho \times c_y$, which means the indirect effect of the missing confounder C was included in the direct effect. The other panels in Figure 6.3 show the sensitivity of identifying a significant direct effect since the true direct effects are all positive. Again, since the estimated direct effect is centered at $c' + \rho \times c_y$ but not c', the sensitivity is high when $c' + \rho \times c_y$ is away from 0 but not when c' is away from 0.

Lastly, Figure 6.4 gives the variances of the direct effect estimate when parameters change. The variance is high when c_y is large. The larger the c_y, the larger the proportion of variances in y cannot be explained by predictors when C is missing from the model. Therefore, the estimated variances are larger.

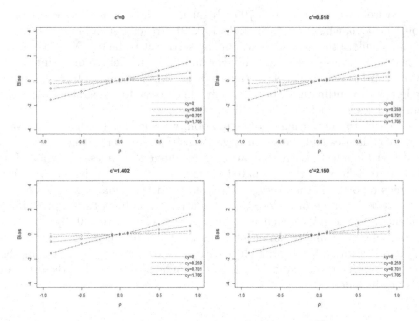

FIGURE 6.2
The bias of the estimated direct effect when Assumption 1 is violated.

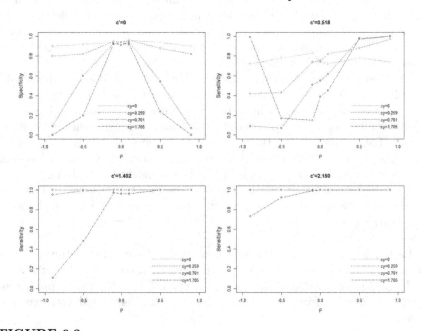

FIGURE 6.3
The sensitivity/specificity of identifying important direct effects when Assumption 1 is violated.

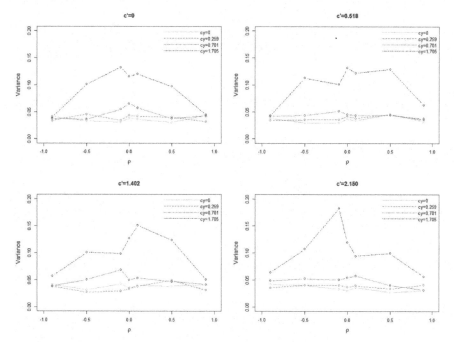

FIGURE 6.4
The variance of direct effect estimates when Assumption 1 is violated.

6.1.2 On the Indirect Effect of M

Using the linear models, the bias of the estimated indirect effect through M is shown in Figure 6.5. We found that no matter what is the true c', the biases of the estimates are all close to 0. The biases do not change with ρ or c_y either.

In Figure 6.6, all indirect effects are at the same level: $a \times b = 0.5419 \neq 0$. Therefore, Figure 6.6 gives the sensitivities of identifying the important third-variable M when parameters change. In general, we find that the sensitivity does not change much when ρ changes but is a little lower when c_y is very high. This is due to that the variance of estimating b becomes higher when an important confounder is missing in the model. When c_y is higher, C contributes more significantly in predicting y.

Figure 6.7 gives the variances of the indirect effect estimate when parameters change. As expected, the variance is higher when c_y is larger.

6.1.3 On the Total Effect

The bias of the estimated total effect is shown in Figure 6.8. Since the total effect is the summation of direct and indirect effect, we found that the biases of the estimated total effects are mainly from the estimated direct effect.

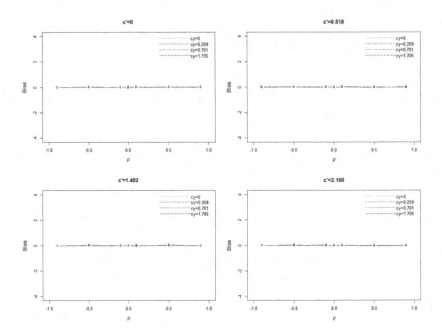

FIGURE 6.5
The bias of the estimated indirect effect of M when Assumption 1 is violated.

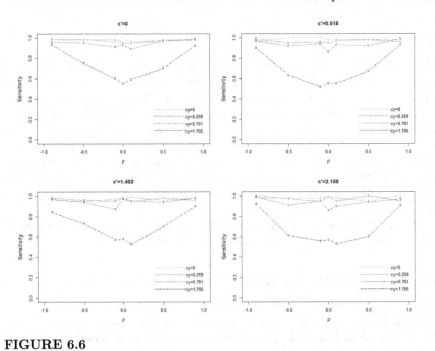

FIGURE 6.6
The sensitivity of identifying the important third-variable M when Assumption 1 is violated.

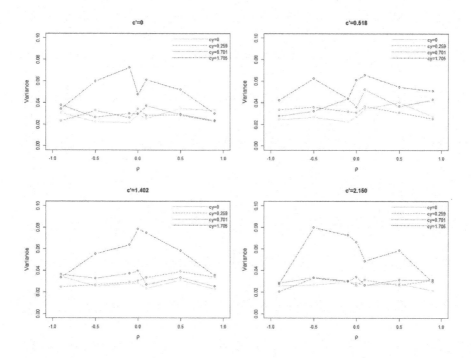

FIGURE 6.7
The variance of indirect effect estimates when Assumption 1 is violated.

Therefore, Figures 6.2 and 6.8 have the same trend. The indirect effect of the confounder C was included in the total effect.

The specificity and sensitivity of the estimated total effect are similar to the estimated direct effect. The true total effect is $c' + a \times b > 0$ for all the selected c's. Since the estimated total effect is centered at $c' + a \times b + \rho \times c_y$, the sensitivity of identifying significant total effect is high when this value is away from 0 but not when the true total effect is away from 0.

Finally, we estimate the variances of the total effect estimates when parameters change. Again, the variance is high when c_y is high. To avoid too many figures in the book, the figures for sensitivity of identifying important total effects and the estimated variances for total effects when Assumption 1 is violated are provided in the book website.

6.1.4 Summary and the Correct Model

In summary, if Assumption 1 is violated, that is, there are unobserved confounders of the $X - Y$ relationship, the effect of the confounders is likely to be mixed into the estimated direct effect. The estimated direct effect can be biased. In the simulated case, the unobserved confounder has very little

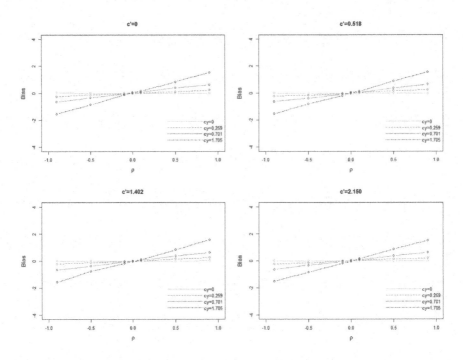

FIGURE 6.8
The bias of the estimated total effect when Assumption 1 is violated.

influence on the estimation of the indirect effect. However, when there are unobserved confounders, the estimated variances of all estimates are likely to be higher.

If the confounder C_i is observed, it should be included in the third-variable analysis. If the purpose is to find out all factors that explain the $X - Y$ relationship, C can be set as a third-variable in the argument *mediator* as well as M. In such case, the total effect is $c' + a \times b + \rho \times c_y$, the direct effect is c', the indirect effect from M is $a \times b$ and from C is $\rho \times c_y$. The estimates of all the third-variable effects are unbiased through the "mma" package.

If the purpose is to find out all causal factors that intervene the $X - Y$ relationship, C should be included as a covariate. In the "mma" package, C and M are included in the data frame argument x but only M is set as a *mediator*. In such case, the total effect does not include the effect from the path *exposure* $- C \rightarrow$ *outcome* and therefore is $c' + a \times b$. The direct effect and indirect effect from M is as above, but the effect through C would not be estimated. The codes for fitting the model for the two situations are included in the book website.

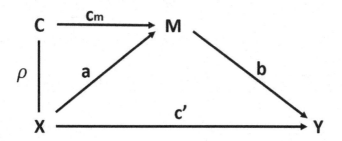

FIGURE 6.9
When the confounder C of the X-M relationship is not observed.

6.2 Assumption 2: No-Unmeasured-Confounder for the Exposure-Third Variable Relationship

The second assumption requires that all confounders for the exposure-third variable relationship are measured and put in the model appropriately. Figure 6.9 shows the relationship among four variables X, M, C and Y, where Y is the outcome and X is the exposure variable. Again, a line with an arrow identifies a causal relationship between the two variables and a line without an arrow means that the two connected variables are related but no causal relationship exists. In the figure, M is a mediator by definition and C is a confounder for the $X - M$ relationship. By Assumption 2, all confounders for the exposure-third variable relationship should be included in analysis. If this is not the case, the estimation of third-variable effects can be influenced.

In Figure 6.9, if C is not included in the analysis, we would like to check how it influences the estimations of direct and indirect effects. Data are generated in the following way:

$$\begin{pmatrix} X_i \\ C_i \end{pmatrix} \sim Bivariate\ Normal \left(\begin{pmatrix} 0 \\ 0 \end{pmatrix}, \begin{pmatrix} 1 & \rho \\ \rho & 1 \end{pmatrix} \right) \tag{6.4}$$

$$M_i = \alpha_1 + aX_i + c_mC_i + \epsilon_{1i} \tag{6.5}$$

$$Y_i = \alpha_2 + bM_i + c'X_i + \epsilon_{2i} \tag{6.6}$$

In the equations, $i = 1, \ldots, 50$, ϵ_{1i} and ϵ_{2i} are independently standard normal distributed. α_1 and α_2 are the intercepts taking the values 0.2 and 0.1, respectively. The correlation coefficient between X and C is ρ, setting at each value from the set $(-0.9, -0.5, -0.1, 0, 0.1, 0.5, 0.9)$. We are interested to see how the different values of ρ can influence the estimation results. c' is the true direct effect of X in this case and is chosen to be 1.402. c_m is selected permuted from $(0, 0.259, 0.701, 1.705)$. The true indirect effect of M should be $a \times b$, where a is chosen from the set $(0, 0.286, 0.773, 1.185)$ and b is 0.701.

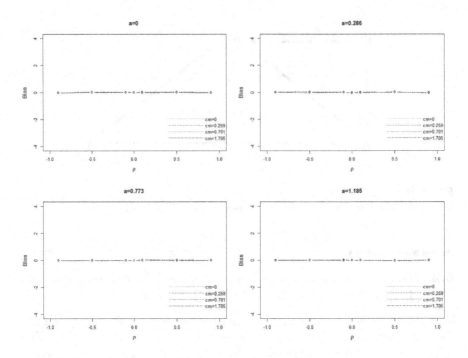

FIGURE 6.10
The bias of the estimated direct effect when Assumption 2 is violated.

Again when C is not included in analysis, we check the performance of estimation using the biases and variances of estimates and the specificity and sensitivity of identifying significant third-variable effects. The estimates are evaluated for the direct effect, indirect effect and total effect, respectively. Each simulation is repeated 100 times.

6.2.1   On the Direct Effect

Using the linear models, the bias of the estimated direct effect is shown in Figure 6.10. We found that the biases of the direct effect estimates are very close to 0 and do not change with parameters.

Next, we check the sensitivity of identifying significant direct effects. Since $c' = 1.402 \neq 0$ for all scenarios, the direct effect is different from 0. We found that the sensitivities are all close to 1.

Lastly, we check the variances of the direct effect estimates when parameters change. We do not find any pattern of the variance changing with parameters. Figures for the variances and sensitivities are provided in the book website.

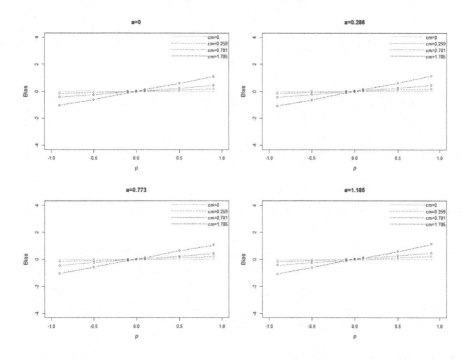

FIGURE 6.11
The bias of the estimated indirect effect of M when Assumption 2 is violated.

6.2.2 On the Indirect Effect of M

Using the linear models with mma, the bias of the estimated indirect effect through M is shown in Figure 6.11. We found that the bias does not change with a. However, the bias of the estimates increases with ρ and with c_m. The mean biases are around $\rho \times c_m \times b$. This means the effect from the exposure-third variable confounder is mixed with the indirect effect of the third-variable when the confounder is not included in analysis.

In Figure 6.12, since the true indirect effect is $a \times b$. The upper left panel of Figure 6.12 gives the specificity of identifying an unimportant third-variable M when it is not significant ($a = 0$). The specificity moves away from 1 when ρ moves away from 0 and when c_m moves away from 0. The other three panels of Figure 6.12 show the sensitivity of finding the significant third-variable M. Since the estimated indirect effect is centered at $(\rho \times c_m + a) \times b$ rather than at the true indirect effect of $a \times b$, the sensitivity moves closer to 1 when the estimated value moves away from 0.

Figure 6.13 gives the variances of the indirect effect estimates when parameters change. The variance is higher when c_m is higher since a bigger c_m is related with a higher confounding effect of C between X and M, so the

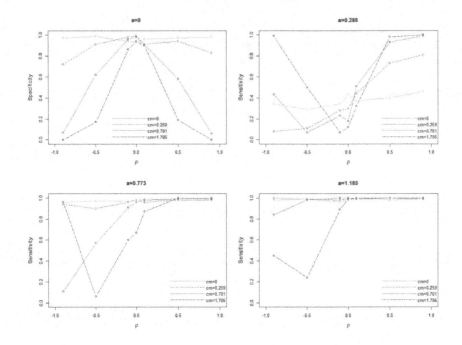

FIGURE 6.12
The sensitivity of identifying the important third-variable M when Assumption 2 is violated.

variance in estimating a is higher. For the same reason, when a is big, the variance increases with ρ.

6.2.3   On the Total Effect

The biases of the estimated total effects have the same trend as the estimated indirect effects. This is because the direct effect estimates are unbiased for this case and the total effect is basically the sum of direct effect and indirect effect. The figure of biases of the estimated total effects is provided in the book website.

Figure 6.14 gives the sensitivities of estimating the important total effect when parameters change. It is the sensitivity since the true total effect is $c' + a \times b > 0$ for all the selected parameters. The sensitivity of identifying significant total effect is a little low when c_m is large and ρ is negative since the estimated total effect is around $c' + (\rho \times c_m + a) \times b$. When ρ is negative and c_m is big, the estimated total effect is dragged toward 0.

Figure 6.15 shows the variances of the total effect estimates when parameters change. The variance is high when c_m is large since C, the important confounder between $X - M$ is ignored in the analysis.

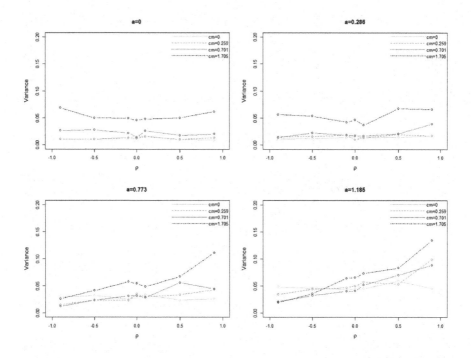

FIGURE 6.13
The variance of indirect effect estimates when Assumption 2 is violated.

6.2.4   Summary and the Correct Model

In summary, when a confounder for the exposure-third variable relationship is unmeasured, the effect from the missed confounder is mixed into the estimated indirect effect, hence resulting in biased estimates for the indirect effect. In the simulation, the missing confounder has little influence on the estimation of the direct effect.

If the confounder C_i is observed, it should be included in the third-variable analysis as a co-variate for predicting M by setting the argument *cova* = C. The codes of including the confounder as a covariate of estimating M in the third-variable analysis are included in the book website.

6.3   Assumption 3: No-Unmeasured-Confounder for the Third Variable-Outcome Relationship

The third assumption requires that all confounders for the third variable-outcome relationship are measured and put in the model appropriately.

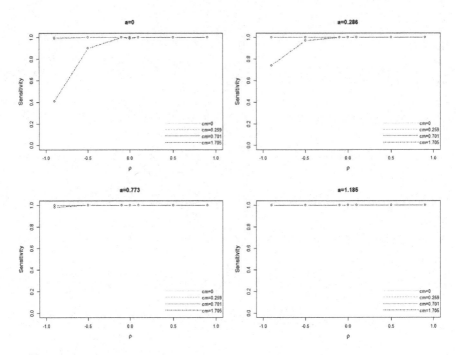

FIGURE 6.14

The sensitivity of identifying important total effects when Assumption 2 is violated.

Figure 6.16 shows the relationship among four variables X, M, C and Y, where Y is the outcome and X is the exposure variable. In the figure, M is a mediator by definition and C is a confounder for the $M - Y$ relationship. By Assumption 3, all confounders for the third variable-outcome relationship should be included in analysis. If this is not the case, the estimation of third-variable effects can be biased.

If C is not included in the analysis, we would like to check how it influences the estimations of direct and indirect effects. The data are generated in the following way:

$$\begin{pmatrix} C_i \\ \epsilon_{1i} \end{pmatrix} \sim Bivariate\ Normal \left(\begin{pmatrix} 0 \\ 0 \end{pmatrix}, \begin{pmatrix} 1 & \rho \\ \rho & 1 \end{pmatrix} \right) \tag{6.7}$$

$$X_i \sim N(0,1) \tag{6.8}$$

$$M_i = \alpha_1 + aX_i + \epsilon_{1i} \tag{6.9}$$

$$Y_i = \alpha_2 + bM_i + c'X_i + c_yC_i + \epsilon_{2i} \tag{6.10}$$

In the equations, $i = 1, \ldots, 50$, ϵ_{2i}s are independently standard normal distributed. α_1 and α_2 are the intercepts taking the values 0.2 and 0.1. The correlation coefficient between C and ϵ_1 is ρ, setting at a value from the set $(-0.9, -0.5, -0.1, 0, 0.1, 0.5, 0.9)$. We are interested to see how the different

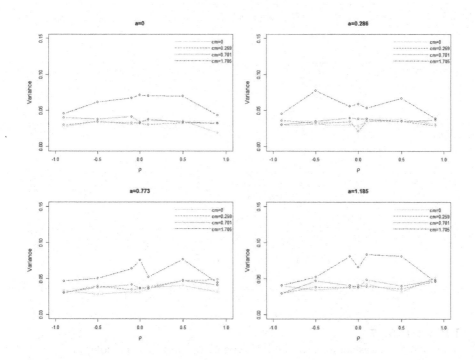

FIGURE 6.15
The variance of total effect estimates when Assumption 2 is violated.

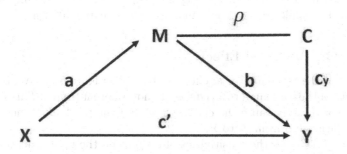

FIGURE 6.16
When the confounder C of the M-Y relationship is not observed.

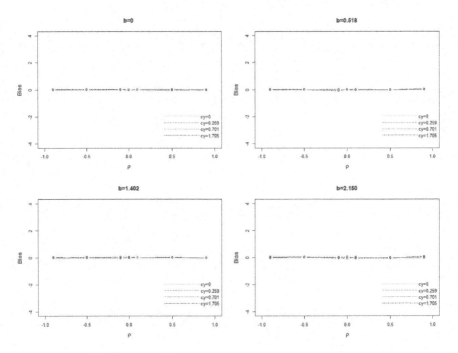

FIGURE 6.17
The bias of the estimated total effect when Assumption 3 is violated.

values of ρ influence the estimation results. c' is the direct effect of X in this case and is set at 0.701. c_y takes a value from the set $(0, 0.259, 0.701, 1.705)$. The indirect effect of M should be $a \times b$, where a is set at 0.773 and b is chosen from the set $c(0, 0.259, 0.701, 1.705)$.

Again, we check the performance of estimation using the biases and variances of estimates and the specificity and sensitivity of identifying the significant third-variable effects. Each simulation is repeated 100 times.

6.3.1   On the Total Effect

The bias of the estimated total effect is shown in Figure 6.17. We found that the biases of the estimated total effects are not different from 0. This is because X does not directly cause the change in C and the total effect measures the direct relationship from X to Y.

Figure 6.18 gives the sensitivities of estimating the important total effect when parameters change. It is the sensitivity since the true total effect is $c' + a \times b > 0$ for all the selected parameters. The estimated total effect is centered at $c'+ab$. The sensitivity of identifying significant total effect is a little lower when c_y increases. This is due to that the variances of estimating the total effects are higher when ρ and c_y are away from 0 as shown in Figure 6.19.

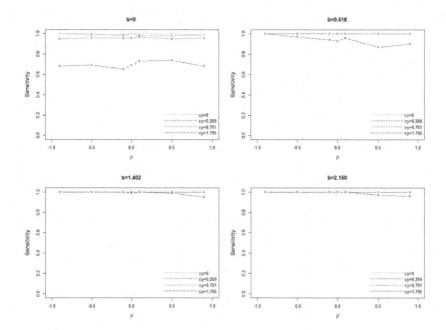

FIGURE 6.18
The sensitivity of identifying important total effects when Assumption 3 is violated.

6.3.2 On the Direct Effect

Using the linear models, the bias of the estimated direct effect is shown in Figure 6.20. We found that the bias of the direct effect increases with c_y but decreases with ρ. This means that ignoring a confounder of the $M - Y$ relationship influences the estimation of the direct effect and the bias is proportional to $\rho \times c_y$.

Next, we check the sensitivity of identifying significant direct effects. In Figure 6.21, since $c' = 0.701 \neq 0$, the direct effect is different from 0. We found that the sensitivity is getting closer to 1 when $c' + \rho \times c_y$ moves away from 0.

Lastly, Figure 6.22 gives the variances of the direct effect estimate when parameters change. The variance does not change much when b changes but it is higher when c_y is higher. This is because high c_y means an important confounder is missed from the linear regression. When ρ is different from 0, the variance decreases. This might be due to that bigger $|\rho|$ means higher correlation between M and C, therefore more variance goes to the estimates of b rather than of c'.

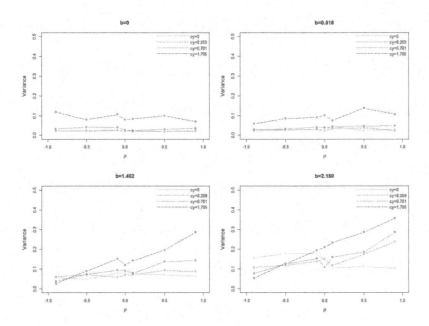

FIGURE 6.19
The variance of total effect estimates when Assumption 3 is violated.

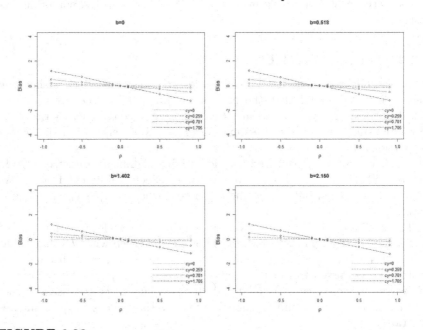

FIGURE 6.20
The bias of the estimated direct effect when Assumption 3 is violated.

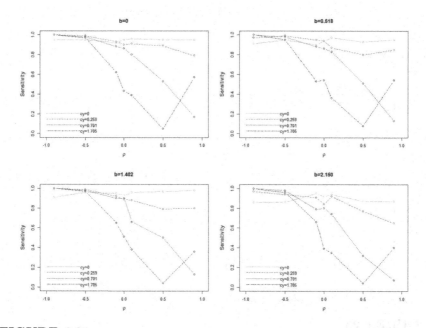

FIGURE 6.21
The sensitivity of identifying important direct effects when Assumption 3 is
violated.

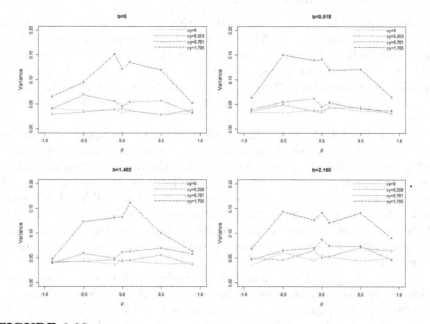

FIGURE 6.22
The variance of direct effect estimates when Assumption 3 is violated.

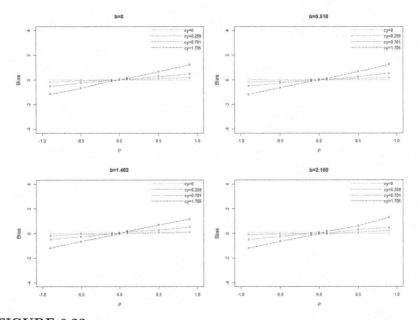

FIGURE 6.23
The bias of the estimated indirect effect of M when Assumption 3 is violated.

6.3.3   On the Indirect Effect of M

Using the linear models, the bias of the estimated indirect effect through M is shown in Figure 6.23. We found that the bias does not change with b. In addition, the bias of the estimates increases with ρ and c_y. The biases for the indirect effect approximately equal the biases for the direct effect in magnitude but are of opposite signs. This is due to that the total effect estimates are unbiased. Therefore the bias of indirect effect and direct effect offset each other.

In Figure 6.24, the true indirect effect is $a \times b$. The upper left panel of Figure 6.24 gives the specificity of identifying an unimportant third-variable M when M is not significant ($b = 0$). Since the bias of the indirect effect estimate is proportional to $-\rho \times c_y$, the specificity is close to 1 when $c_y = 0$ or $\rho = 0$, but decreases when c_m or ρ moves away from 0. When $b \neq 0$, since $a = 0.701 \neq 0$, the indirect effect of M is not 0. The other three panels of Figure 6.24 show the sensitivity of identifying the significant M. we find that the sensitivity increases when the estimated indirect effect, $a \times (b + \rho c_y)$ moves away from 0.

Figure 6.25 gives the variances of the indirect effect estimate when parameters change. The variance is higher when c_y is higher since a bigger c_y is related with a higher confounding effect of C between X and M, so the variance in estimating b is higher. In addition, we found that when b is big, the variance increases with ρ.

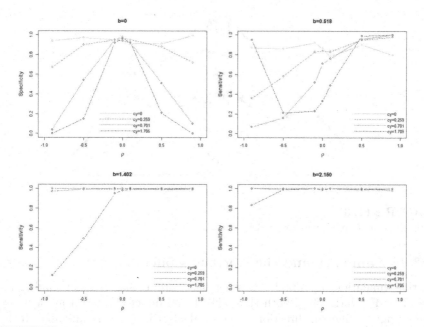

FIGURE 6.24
The sensitivity/specificity of identifying the important third-variable M when Assumption 3 is violated.

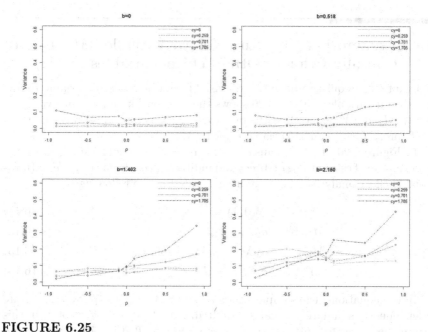

FIGURE 6.25
The variance of indirect effect estimates when Assumption 3 is violated.

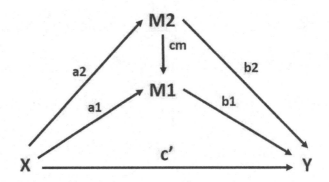

FIGURE 6.26
When M_2 is causally prior to M_1.

6.3.4 Summary and the Correct Model

In this simulation, we found that when there are unobserved confounders for the $M - Y$ relationship, both the indirect and direct effects estimations are biased but to different directions. The total effect is therefore unbiased. If the confounder C_i is observed, it should be included in the third-variable analysis as a co-variate. The codes for the right model are included in the Appendix.

6.4 Assumption 4: Any Third-Variable M_i is not Causally Prior to Other Third-Variables in $\mathbf{M}_{-i}$.

Assumption 4 requires that no third-variable can be a causal variable to any other third-variables. Figure 6.26 shows the relationship among four variables X, M_1, M_2 and Y, where Y is the outcome and X is the exposure variable. In the figure, M_2 is a causal variable prior to M_1.

In Figure 6.26, if M_2 is causally prior to M_1, we would like to check the performance of estimations of direct and indirect effects from the general third-variable effect analysis. The data are generated in the following way:

$$X_i \sim N(0,1) \tag{6.11}$$
$$M_{2i} = a_2 X_i + \epsilon_{1i} \tag{6.12}$$
$$M_{1i} = a_1 X_i + c_m M_{2i} + \epsilon_{2i} \tag{6.13}$$
$$Y_i = c' X_i + b_1 M_{1i} + b_2 M_{2i} + \epsilon_{3i} \tag{6.14}$$

In the equations, the sample size is 50 so $i = 1, \ldots, 50$, $\epsilon_{1i}, \epsilon_{2i}$ and ϵ_{3i} are independently standard normal distributed. c' is the direct effect of X in this case and is chosen to be 1.25. b_2 ranges from $(0, 0.259, 0.701, 1.705)$ and a_2 from $(0, 0.286, 0.773, 1.185)$. c_m is chosen from $(-0.9, -0.5, -0.1, 0, 0.1, 0.5, 0.9)$. a_1

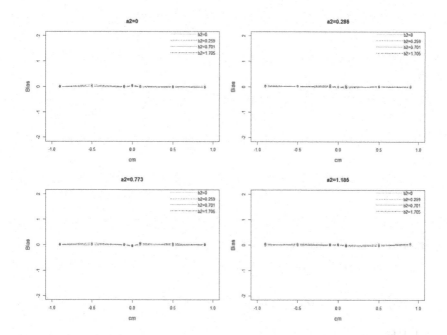

FIGURE 6.27
The bias of the estimated direct effect when Assumption 4 is violated.

is set at 0.773 and b_1 at 0.701. The true indirect effect from M_2 is $a_2 \times b_2$, and the effect from M_1 is $a_1 \times b_1 + a_2 \times c_m \times b_1$.

We check the performance of estimation of the direct effect, indirect effects of each third-variable and the joint indirect effect of both third-variables. Each simulation is repeated 100 times.

6.4.1 On the Direct Effect

Using the linear models, the bias of the estimated direct effect is shown in Figure 6.27. We found that no matter how parameters change, the biases of the estimated direct effect are close to 0. There is no significant bias in estimating the direct effect.

Next, we check the sensitivity of identifying significant direct effects. In Figure 6.28, since $c' = 1.25$, the sensitivity is very close to 1 and does not change with a_2, b_2 or c_m.

Lastly, Figure 6.29 shows the variances of the direct effect estimate when parameters change. In summary, the variances do not change with parameters either.

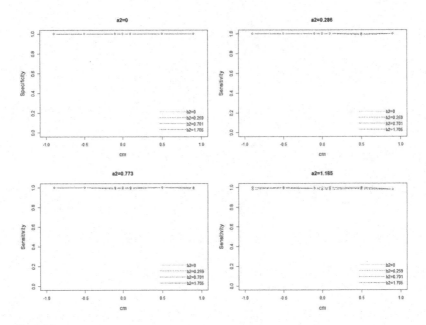

FIGURE 6.28
The sensitivity/specificity of identifying important direct effects when Assumption 4 is violated.

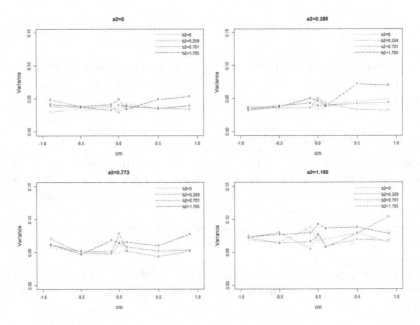

FIGURE 6.29
The variance of direct effect estimates when Assumption 4 is violated.

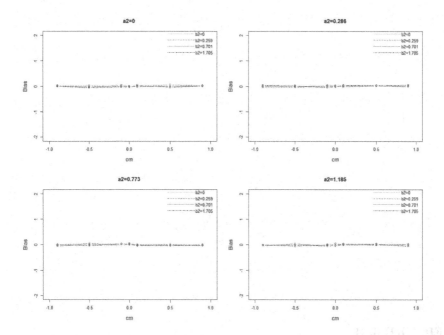

FIGURE 6.30
The bias of the estimated indirect effect of M_1 and M_2 when Assumption 4 is violated.

6.4.2 On the Indirect Effect of M

Using the linear models, the biases of the estimated indirect effect through M_1 and M_2, respectively are close to 0 for all parameters. Therefore, the estimated joint effect of M_1 and M_2 is unbiased as is shown in Figure 6.30.

In Figure 6.31, the indirect effect of M_1 is $a_1 \times b_1 + a_2 \times c_m \times b_1$, the sensitivity/1-specificity of identifying M_1 as an important third-variable is closer to 1 when the true indirect effect moves away from 0. Figure 6.32 gives the sensitivities and specificity of identifying the important third-variable M_2 when parameters change. In general, the upper left panel of Figure 6.32 is the specificity of identifying M_2 since $a2 = 0$, we find the specificity is very close to 1. The other three panels are the sensitivity of finding M_2 as an important third-variable when $b_2 \neq 0$ and is $1 - specificity$ when $b_2 = 0$. The sensitivity/1 - specificity increases with a_2 or b_2 but does not change much with c_m except that the sensitivity is a little lower at both ends of c_m. This is because the variances of the indirect effects are a little higher, as is shown by Figure 6.35. The sensitivity/1-specifity of the joint effect of M_1 and M_2 is generally closer to 1 when the true joint effect moves away from 0 as is shown in Figure 6.33.

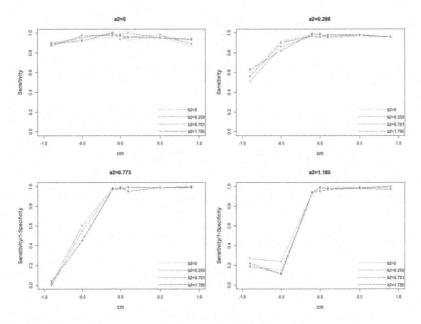

FIGURE 6.31
The sensitivity and specificity of identifying the important third-variable M_1 when Assumption 4 is violated.

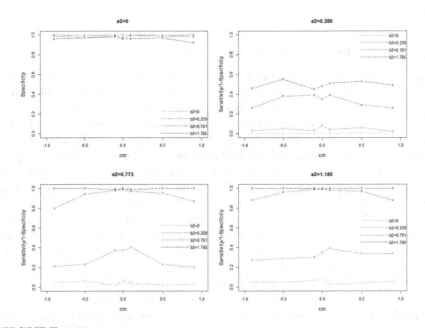

FIGURE 6.32
The sensitivity and specificity of identifying the important third-variable M_2 when Assumption 4 is violated.

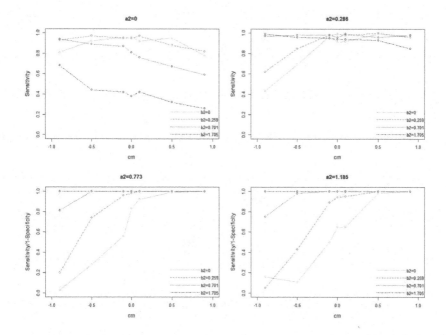

FIGURE 6.33
The sensitivity and specificity of identifying the joint third-variable effects of M_1 and M_2 when Assumption 4 is violated.

Figures 6.34, 6.35 and 6.36 give the variances of the indirect effect estimates when parameters change. The variance for the estimated ie for M_1 increases with c_m and a_2, but not with b_2. The variance for the estimated ie for M_2 increases with b_2 and a_2. By Delta method, the variance is close to $\hat{a}_2^2\widehat{var}(b_2) + \hat{b}_2^2\widehat{var}(a_2)$. The variances of the joint effect estimates have the same tendencies.

6.4.3 On the Total Effect

The bias of the estimated total effect is shown in Figure 6.37. We found that the biases of the estimated total effects are close to 0 and do not change with parameters.

The sensitivities of estimating the total effect are all close to 1. Figure 6.38 gives the variances of the total effect estimate when parameters change. The variance is about the same with a_2 but is higher when c_m or b_2 is higher.

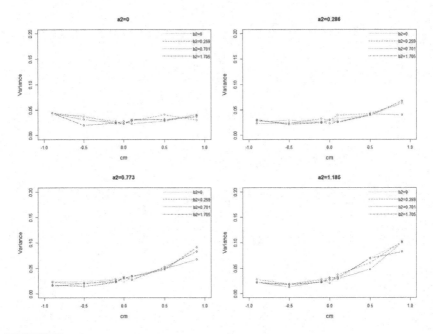

FIGURE 6.34
The variance of indirect effect from M_1 estimates when Assumption 4 is violated.

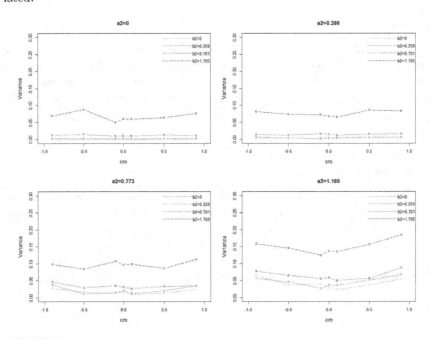

FIGURE 6.35
The variance of indirect effect from M_2 estimates when Assumption 4 is violated.

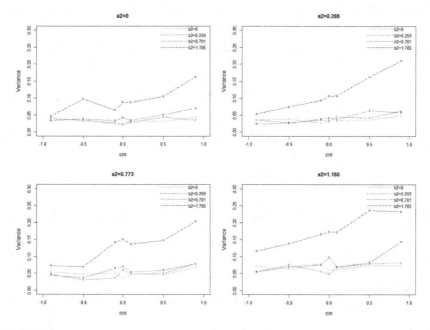

FIGURE 6.36
The variance of the joint indirect effect estimates of M_1 and M_2 when Assumption 4 is violated.

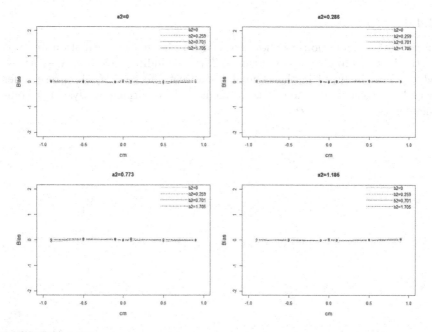

FIGURE 6.37
The bias of the estimated total effect when Assumption 4 is violated.

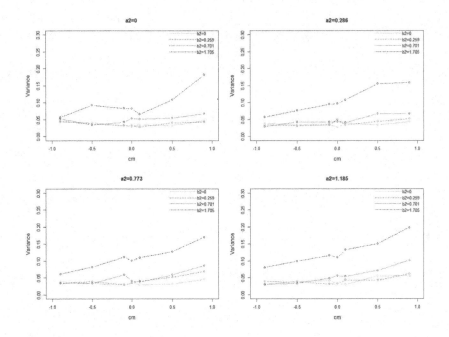

FIGURE 6.38
The variance of total effect estimates when Assumption 4 is violated.

6.4.4   Conclusion

In general, Assumption 4 is not required for the general mediation analysis method. When M_2 is causally prior to M_1. The indirect effect of $X \to M_2 \to M_1 \to Y$ is included in the estimated indirect effect of M_1. If a researcher would like to separate the effects, sequential third-variable analysis is recommended (see Chapter 12).

7

Multiple Exposures and Multivariate Responses

In explaining health related disparities, we often need to deal with multi-categorical or multiple exposures and/or multivariate outcomes. A motivating example is that we want to jointly consider the racial and ethnic disparities that exist in both the body mass index (BMI) and in the prevalence of obesity. In this chapter, we extend the general TVEA to deal with multivariate exposures/outcomes. We also include a method for finding joint confidence intervals (confidence ball) for all estimated third-variable effects. This chapter is derived in part from an article published in the Journal of Applied Statistics [82] on March 8th, 2020, available online: http://www.tandfonline.com/10.1080/02664763.2020.1738359.

7.1 Multivariate Multiple TVEA

The framework of the multivariate TVEA is shown in Figure 7.1. In the plot, $X_1, \ldots, X_S$ are the exposure variables. $M_1, \ldots, M_p$ are potential third-variables. Y is the outcome variable. It can be a vector or a scalar. Z is the vector of other covariates that relate with Y but not with X. A line between two variables means that the two variables are associated (causally for mediation effect or not-causally for confounding effect). A path is defined by the line(s) connecting each exposure variable X with each of the outcome Y. In our motivating example of jointly exploring the racial and ethnic disparities in both BMI and obesity, Y is a vector with Y_1 being the BMI of the subjects, which is continuous and Y_2 being the indicator of obese, a binary variable. The exposure variable has three levels: Non-Hispanic White (NHW), Hispanic White (HW) and Non-Hispanic Black (NHB). Therefore we have two exposures, X_1 (HW or not) and X_2 (NHB or not), which are both binary.

We propose algorithms of using general predictive models to make inferences on the third-variable effects where the multivariate exposure variables can be both binary and/or continuous. Note that a multi-categorical exposure of K categories can be decomposed into $K - 1$ binary exposures. Our purpose is to make inferences on the set of all direct and indirect effects for each pair

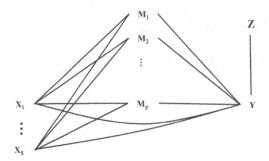

FIGURE 7.1
The framework for multicategorical/multivariate exposure TVEA. $X_1, \ldots, X_s$ are the exposure variables, $M_1, \ldots, M_p$ are potential MCs. Y is the outcome variable (vector), and Z is the vector of other covariates that relate with Y but not X. A line between two variables means that the two variables are associated (causally or not-causally).

of exposure-outcome variable relationship if there is at least one path connecting the exposure-outcome pair. We present the algorithms for estimations of third-variable effects when the exposures are multi-categorical. For the continuous or mixed exposures, the algorithms can be easily extended and have been implemented in the *mma* R package we created.

7.1.1 Non/Semi-Parametric TVEA for Multi-Categorical Exposures

To handle multicategorical exposure variables, Algorithms 5 and 6 are derived directly from the definitions of third-variable effects in Chapter 4. The multicategorical exposure variable is transformed into multiple binary variables. The Algorithms provide a method to calculate third-variable effects with multiple binary exposure variables. The algorithms proposed here are implemented in the R package *mma*.

The algorithms are based on the predictive model for $E(Y)$, which has the form

$$E(Y_i) = f(X_{1i}, \ldots, X_{Si}, M_{1i}, \ldots, M_{pi}, Z_i), \quad \text{for } i = 1, \ldots, n,$$

(7.1)

where $X_{1i}, \ldots, X_{Si}$ are the multivariate binary exposure variables. X_{si} is 0 or 1. There is at most one of the $X_{si}, s = 1, \ldots, S$ can be 1 for each i. In the R package *mma*, MART, Cox hazard function, and generalized linear models (GLM) can be used to build f. A researcher can also build up his/her own predictive functions through the *custom.function* argument. Note that the outcome can be a vector, in which case f is a vector of functions. In the algorithms, the reference group is the group when all X_s are zero, and the sth group is identified when $X_s = 1$. Also, n_s denotes the size of the *sth* group in the original data set, where $s = 0, \ldots, S$.

Algorithm 5 *The total effect for exposure variable X_s, denoted as TE_s, where $s = 1, \ldots, S$, is the average difference in Y between the sth group and the reference group. To estimate TE_s:*

1. Randomly draw n_0 vectors of third-variables and covariates from the reference group, where all $X_k = 0, k = 1, \ldots, S$. Denote the sampled vectors as $(M_{1j1}, \ldots, M_{pj1}, Z_{j1})^T$, for $j = 1, \ldots, n_0$.

2. Randomly draw n_s vectors of third-variables and covariates from the sub-population where $X_s = 1$. Denote the sampled vectors as $(M_{1j2}, \ldots, M_{pj2}, Z_{j2})^T$, for $j = 1, \ldots, n_s$.

3. $TE_s = \frac{1}{n_s} \sum_{j=1}^{n_s} f(0, \ldots, X_{sj2} = 1, \ldots, 0, M_{1j2}, \ldots, M_{pj2}, Z_{j2}) - \frac{1}{n_0} \sum_{j=1}^{n_0} f(0, \ldots, 0, M_{1j1}, \ldots, M_{pj1}, Z_{j1})$.

We may also calculate the total effect using $E(Y|X_s = 1) - E(Y|X = 0_S)$ directly from the observations.

Note that when any third-variables depend on a set of covariates in addition to the exposure variables, the sample $(M_{1j1}, \ldots, M_{pj1}, Z_{j1})^T$ is drawn from the conditional distribution given both $X_k = 0, k = 1, \ldots, S$ and the set of covariates. Also the sample $(M_{1j2}, \ldots, M_{pj2}, Z_{j2})^T$ is drawn from the conditional distribution given both $X_s = 1$ and the covariates. In such cases, multivairate models are required to fit the relationship between each third-variables and the exposures and covariates. The samples are to be used again in Algorithm 6 Step 3.

Algorithm 6 *The direct effect not through M_k for exposure variable X_s, denoted as $DE_{s \backslash M_k}$, where $s = 1, \ldots, S$, is the average difference in Y between the sth group and the reference group when the relationship between X and M_k is broken. To estimate $DE_{s \backslash M_k}$:*

1. Use the samples generated by Steps 1 and 2 of Algorithm 1.

2. Combine the vectors $\{M_{kj1}\}_{j=1}^{n_0}$ and $\{M_{kj2}\}_{j=1}^{n_s}$ and randomly permute the combined vector, denote the new vector as $\{\tilde{M}_{kj}\}_{j=1}^{n_0+n_s}$. $\{\tilde{M}_{kj}\}_{j=1}^{n_0+n_s}$ forms a sample of M_k from the conditional distribution such that all $X_l = 0$, where $l \neq s$.

3. $DE_{s \backslash M_k}$ is estimated by $\frac{1}{n_s} \sum_{j=1}^{n_s} f(0, \ldots, X_{sj2} = 1, \ldots, 0, M_{1j2}, \ldots, M_{k-1,j2}, \tilde{M}_{kj}, M_{k+1,j2}, \ldots, M_{pj2}, Z_{j2}) - \frac{1}{n_0} \sum_{j=1}^{n_0} f(0, \ldots, 0, M_{1j1}, \ldots, M_{k-1,j1}, \tilde{M}_{k,(n_s+j)}, M_{k+1,j1}, \ldots, M_{pj1}, Z_{j1})$.

Due to the randomness brought in by sampling, the two algorithms are repeated, and the average results from the repetitions are estimates of the third-variable effects. For the analysis in this chapter, we set the re-sampling times at 20. Also note that for each combination of exposure-outcome relationship, there is a set of third-variable effects estimated. For the motivating example, we explore four sets of third-variable effects: the racial (NHW vs. NHB) and ethnic (NHW vs. HW) disparities in obesity and BMI separately. The four sets of estimates can be highly correlated.

It is important on selecting the reference group for the exposure variable. For example, if we choose HW as the reference group, the comparisons would be between HW and NHW, and HW and NHB. But the comparison between HW and NHB cannot separate either the racial or the ethnic difference. Although, theoretically, if we use HW as the reference group, the third-variable effects comparing NHW with NHB can be estimated by the difference in the estimated third-variable effects when comparing HW with NHW, and comparing HW with NHW.

7.1.2   Non/Semi-Parametric TVEA for Multiple Continuous Exposures

The algorithms to estimate the third-variable effects with multiple continuous exposures are the same as Algorithms 3 and 4 except that Equation 4.4 is changed to Equation 7.2 as in the following:

$$
\left(\begin{array}{c|c} M_{1i} \\ M_{2i} \\ \ldots \\ M_{pi} \end{array} \, \mathbf{x}_i, \mathbf{cm}_i \right) \sim \mathbf{\Pi} \left(\left(\begin{array}{c} g_1(x_i) \\ g_2(x_i) \\ \ldots \\ g_p(x_i) \end{array} \right), \mathbf{\Sigma} \right) \tag{7.2}
$$

In Equation 7.2, the conditional distribution of third-variables is based on the vector of exposure variables, $\mathbf{X}$, and the covariates for third-variables, $\mathbf{CM}$. $\mathbf{\Pi}$ is the joint distribution of $\mathbf{M}$ given $\mathbf{X}$ and $\mathbf{CM}$, which has a mean vector $\mathbf{g}(x_i)$ and variance-covariance matrix $\mathbf{\Sigma}$.

7.2   Confidence Ball for Estimated Mediation Effects

To estimate the variances and confidence intervals of the estimated third-variable effects, we use the bootstrap method [16]. The typical parametric way of getting the confidence interval is to assume a normal distribution for the estimates and then calculate standard deviations of estimates and build up confidence intervals. A usual non-parametric/empirical method to build $(1-\alpha) \times 100\%$ confidence intervals for estimates is to find the $\frac{\alpha}{2}th$ and $(1-\frac{\alpha}{2})th$ quantiles of the bootstrap estimates for each effect and use them as the lower and upper bounds of the confidence interval.

When there are multiple estimates, the confidence intervals calculated in the above ways are significantly narrower than the truth. The overall type I error is inflated since multiple comparisons are performed. To handle the problem, we introduce the concept of confidence ball. Our purpose is to find out a joint confidence interval for all estimates at the $(1 - \alpha)$ confidence level.

Let $\theta = (\theta_0, \theta_1, \ldots, \theta_p)^T$ be a set of third-variable effects to explore an exposure-outcome relationship (e.g. racial disparity in obesity), where θ_0 is the direct effect and θ_i is the indirect effect from the ith third-variable. We first develop appropriate point estimates for all third-variable effects in the set. Usually the point estimates are the estimated third-variable effects based on all observations. Denote the set of point estimates as $\widehat{\theta}$. The idea is to find a joint confidence ball, in which the squared distance from the ball surface to the center, $\widehat{\theta}$, is minimized. Assume that there are N sets of bootstrap samples. Denote the set of estimates from the ith bootstrap sample as $\widetilde{\theta}_i = (\widetilde{\theta}_{i0}, \ldots, \widetilde{\theta}_{ip})^T, i = 1, \ldots, N$. To build up the confidence ball, we use the following algorithm:

Algorithm 7 *Estimate the $(1 - \alpha) \times 100\%$ confidence ball for the estimated third-variable effects:*

> *1. Calculate the squared distance from each set of bootstrap estimates to the set of point estimates, $d_i = ||\widetilde{\theta}_i - \widehat{\theta}||^2$, for $i = 1, \ldots, N$, where $|| \cdot ||^2$ is the Euclidean distance from $\cdot$ to the origin.*
>
> *2. Let $Ball = \{\widetilde{\theta}_i : d_i \leq Q_{d_i, 1-\alpha}\}$, where $Q_{d_i, 1-\alpha}$ is the $(1 - \alpha) \times 100\%$ percentile of $d_i, i = 1, \ldots, N$.*
>
> *3. The $(1-\alpha) \times 100\%$ confidence interval for θ_j, is $(\min\{\widetilde{\theta}_{ij}\}, \max\{\widetilde{\theta}_{ij}\})$, for all $\theta_i \in Ball$.*

A confidence ball is found for each set of exposure-outcome relationship.

7.2.1 A Simulation Study to Check the Coverage Probability of the Confidence Ball

The coverage probability of a set of confidence intervals is the proportion of times that the confidence intervals cover all true values. We did a series of simulations to check the coverage probability of the confidence ball method when there are two or three third-variables. The confidence ball is compared with the confidence intervals from the normal approximation, and with those from the nonparametric quintiles. The data were generated in the following ways:

$$
\begin{aligned}
x_i &\sim N(0, 1) \\
m_{ji} &= x_i + \epsilon_{ji} \\
y_i &= x_i + \beta m_{1i} + \beta m_{2i} + \beta m_{3i} + \epsilon_i
\end{aligned}
$$

TABLE 7.1
Comparisons of coverage probabilities.

ρ	$J=2$			$J=3$		
	CB	**CQ**	**NM**	**CB**	**CQ**	**NM**
−0.9	0.94	0.95	0.93	0.87	0.82	0.84
−0.5	0.96	0.90	0.94	0.96	0.87	0.91
−0.1	0.92	0.88	0.83	0.97	0.85	0.94
0.0	0.94	0.87	0.84	0.97	0.85	0.88
0.1	0.98	0.89	0.89	1.00	0.92	0.92
0.5	1.00	0.84	0.93	0.98	0.86	0.91
0.9	0.98	0.91	0.95	0.92	0.84	0.84

Note: CB refers to the proposed confidence ball. CQ refers to the confidence intervals from the quantiles. NM refers to the normal approximation method.

where ϵ_i is the random error term with a standard normal distribution. ϵ_{ji} has a multivariate normal distribution with mean 0, variance 1, and the covariance $cov(x_{j_1,i}, x_{j_2,i}) = \rho$ for $j \in 1, \ldots, J$. It is easy to show from the data generation mechanism that the indirect effect for each third-variable is β. We set $\beta = 0.518$, indicating a medium indirect effect, $i = 100$, and J is 2 or 3. The coverage probabilities with different method to generate the 95% confidence intervals are shown in Table 7.1. We generate each dataset 100 times to get the coverage probabilities.

We found that the confidence ball method (CB) maintained a reasonable coverage probability at around 0.95. In addition, the coverage probability does not reduce with the number of parameters. While using the nonparametric quantile (CB) or normal approximation (NM), the coverage probabilities are generally smaller than 95% and the probabilities reduces with the number of parameters. This is due to that the overall type-I error is not controlled under the CB or NM method.

7.3 The R Package *mma*

A more recent version of the *mma* (published after March 8th, 2018) package includes the implementation of general TVEA with multi-categorical/multivariate exposures and with multivariate responses. It also includes the Algorithm 7 proposed in Section 7.2 to calculate and report confidence balls.

The following is an example of using the "weight-behavior" dataset that is included in the *mma* package for a multivariate multiple third-variable analysis. In the example, the predictor is multicategorical of k levels. By the function *data.org*, the predictor is firstly transformed into $k - 1$ binary predictors. In the following example, the predictor is the race with six levels. As a result of the summary of the output of *data.org*, P-value 2 gives the test results of the relationship between each third-variable with each predictor. If a candidate third-variable significantly relates with any of the $k - 1 = 5$ predictors, the variable is kept for further tests.

R codes:

```
#multivariate predictor
 x=weight_behavior[,c(2:3,5:14)]
 pred=weight_behavior[,4]
 y=weight_behavior[,15]
 data.mb.b <-data.org(x,y,mediator=5:12,jointm=list(n=1,
                 j1=c(5,7,9)),pred=pred,predref=
                 "OTHER",alpha=0.4,alpha2=0.4)
 summary(data.mb.b)
```

Results:

```
Identified as mediators:
[1] "tvhours"   "cellhours" "sweat"      "sports"    "gotosch"
Selected as covariates:
[1] "age"        "sex"        "numpeople" "car"       "exercises"
Tests:
```

	P-Value 1.y	P-Value 2.pred	P-Value 2.predAFRICAN
age	0.836	NA	NA
sex	0.178	NA	NA
numpeople	0.561	NA	NA
car	0.052	NA	NA
gotosch -	0.710	0.112	0.796
snack	0.724	NA	NA
tvhours -	0.830	0.748	0.535
cmpthours	0.826	NA	NA
cellhours -	0.067	0.880	0.994
sports *	0.000	0.587	0.163
exercises	0.176	0.629	0.731
sweat *	0.181	0.647	0.174
pred	0.543	NA	NA
predAFRICAN	0.663	NA	NA
predCAUCASIAN	0.242	NA	NA
predINDIAN	0.890	NA	NA
predMIXED	0.782	NA	NA

	P-Value 2.CAUCASIAN	P-Value 2.INDIAN	P-Value 2.MIXED
age	NA	NA	NA
sex	NA	NA	NA
numpeople	NA	NA	NA
car	NA	NA	NA
gotosch -	0.996	0.097	0.033
snack	NA	NA	NA
tvhours -	0.334	0.916	0.092

```
cmpthours               NA              NA              NA
cellhours -             0.707           0.555           0.383
sports *                0.816           0.453           0.091
exercises               0.723           0.912           0.881
sweat *                 0.214           0.374           0.203
pred                    NA              NA              NA
predAFRICAN             NA              NA              NA
predCAUCASIAN           NA              NA              NA
predINDIAN              NA              NA              NA
predMIXED               NA              NA              NA
----
*:mediator,-:joint mediator
P-Value 1:Type-3 tests in the full model (data.org) when testtype=1,
univariate relationship test with the outcome when testtype=2.
P-Value 2:Tests of relationship with the Predictor
```

Similarly, the package can deal with multivariate outcomes. The following code deals with multiple predictors and multivariate responses. If a third-variable candidate is significantly related with any one of the outcomes, and with any of the multiple predictors, the variable is identified as a third-variable with potential significant indirect effect. The results from *data.org* are summarized for each combination of the exposure-outcome relationship.

R codes:

```
#multivariate responses
 x=weight_behavior[,c(2:3,5:14)]
 pred=weight_behavior[,4]
 y=weight_behavior[,c(1,15)]
 data.mb.mb<-data.org(x,y,mediator=5:12,jointm=list(n=1,
                  j1=c(5,7,9)),pred=pred,predref=
                  "OTHER", alpha=0.4,alpha2=0.4)
 summary(data.mb.mb)
```

Results:

```
Identified as mediators:
[1] "tvhours"    "cellhours" "sweat"      "sports"     "gotosch"
Selected as covariates:
[1] "age"        "sex"        "numpeople" "car"        "exercises"
Tests:
              P-Value 1.bmi P-Value 1.overweigh P-Value 2.pred
age              0.458           0.836              NA
sex              0.003           0.178              NA
numpeople        0.282           0.561              NA
car              0.059           0.052              NA
gotosch -        0.527           0.710              0.112
snack            0.830           0.724              NA
tvhours -        0.505           0.830              0.748
cmpthours        0.676           0.826              NA
cellhours -      0.084           0.067              0.880
```

	P-Value 1	P-Value 2	P-Value 3
sports *	0.001	0.000	0.587
exercises	0.089	0.176	0.629
sweat *	0.078	0.181	0.647
pred	0.344	0.543	NA
predAFRICAN	0.900	0.663	NA
predCAUCASIAN	0.032	0.242	NA
predINDIAN	0.446	0.890	NA
predMIXED	0.833	0.782	NA

	P-Value 2.predAFRICAN	P-Value 2.predCAUCASIAN
age	NA	NA
sex	NA	NA
numpeople	NA	NA
car	NA	NA
gotosch -	0.796	0.996
snack	NA	NA
tvhours -	0.535	0.334
cmpthours	NA	NA
cellhours -	0.994	0.707
sports *	0.163	0.816
exercises	0.731	0.723
sweat *	0.174	0.214
pred	NA	NA
predAFRICAN	NA	NA
predCAUCASIAN	NA	NA
predINDIAN	NA	NA
predMIXED	NA	NA

	P-Value 2.predINDIAN	P-Value 2.predMIXED
age	NA	NA
sex	NA	NA
numpeople	NA	NA
car	NA	NA
gotosch -	0.097	0.033
snack	NA	NA
tvhours -	0.916	0.092
cmpthours	NA	NA
cellhours -	0.555	0.383
sports *	0.453	0.091
exercises	0.912	0.881
sweat *	0.374	0.203
pred	NA	NA
predAFRICAN	NA	NA
predCAUCASIAN	NA	NA
predINDIAN	NA	NA
predMIXED	NA	NA

```
*:mediator,-:joint mediator
P-Value 1:Type-3 tests in the full model (data.org) when testtype=1,
univariate relationship test with the outcome when testtype=2
P-Value 2:Tests of relationship with the Predictor
```

Other generic functions *med, boot.med, summary* and *plot* are extended to handle multiple predictors and/or multivariate outcomes in the same way. The outputs are designed for each pair of predictor-outcome relationships separately.

7.4   Racial and Ethnic Disparities in Obesity and BMI

African and Hispanic Americans have a higher prevalence of obesity and related chronic diseases such as diabetes and hypertension when compared to other races and ethnicity in the United States [51, 37]. Mechanisms explaining these disparities are poorly understood. Both neighborhood and individual-level risk factors such as neighborhood walkability and physical activity behavior, are shown to contribute to these racial/ethnic disparities [52, 58]. The purpose of the study is to differentiate factors that can explain the racial and/or ethnic differences in BMI and obesity prevalence. We use the National Health and Nutrition Examination Survey (NHANES) for the study. This variable description is derived in part from an article published in the Spatial and Spatio-temporal Epidemiology [85].

The NHANES is conducted by the National Center for Health Statistics (NCHS) of the Centers for Disease Control and Prevention (CDC) from 1971 onward, with biennial surveys beginning in 1999. The study includes a survey, a medical exam and a laboratory test that capture an array of health information on various physiologic measures, health outcomes and diseases, and health behaviors. We have used the data set to explore the racial disparity of obesity. Readers are referred to [85] for a detailed description of the NHANES study and the inclusive rules of subjects.

The sources of the neighborhood data are the socio-demographic data from US Census 2000 and the American Community Survey, geographic data from ArcGIS and US Census shapefiles, and food environment information drawn from North American Industry Classification System (NAICS), Standard Industrial Classification (SIC) data from InfoUSA and Environmental Systems Research Institute (ESRI). NAICS/SIC data obtained include business name, geo-coded location, and detailed SIC industry codes for food establishments. Census data include various measures as described below.

Given that the temporal frame of the geo-referenced continuous NHANES data ranges from 1999 through 2010, the project temporally aligned the data collected from different sources. Census derived measures were drawn from the 2000 and 2010 census and linearly interpolated between decennial censuses. NCAIS data were obtained historically at five-year intervals going back to 2000. Estimates for measures between interval points were also linearly interpolated, where possible, informed by the ACS. The exception was the Census shape files. All definitions of neighborhood were based on the Census 2000 TIGER shape file.

7.4.1 Variables

Individual Level Measures: Key measures used from NHANES are categorized below.

- Obesity Related Impact Variables: The primary impact measures for the study were body mass index (BMI), defined as weight in kilograms (kg) divided by height in squared meters (m2), and obesity, defined as BMI larger than 30. Analyses were conducted for obesity status as a binary variable and BMI as a continuous variable.

- Dietary Behavior Variables: Individual dietary variables in NHANES are obtained from a 24-hour dietary recall, as well as a questionnaire on dietary behavior. Specifically, two key dietary factors typically associated with obesity were examined in this study, 1) total energy intake and 2) sugar-sweetened beverage consumption. Both variables were operationalized as tertiles characterizing low, medium, and high categories. Physical Activity Variable: Physical activity was assessed using the accelerometer data available in the 2003–2004 and 2005–2006 cycles of the NHANES. The primary variables of interest were total energy expenditure and the level of physical activity low to none, light, moderate, vigorous defined based on the Metabolic Equivalent Task (MET) method. The physical activity variable was dichotomized as none to light physical activity and moderate to vigorous physical activity.

- Control Variables: Other covariates considered in the analysis at the individual level included age, sex, race/ethnicity, education, family history of disease, language used/spoken at home, type of employment/occupation, income, household size, health insurance (yes/no), tobacco use, and alcohol use.

Neighborhood Level Measures: Neighborhood was defined as census tract of residence.

- Food Environment Variables: We defined and examined the impact of the neighborhood food environment as the density of specific types of food establishments (e.g. outlets per capita) using a continuous scale. Types of food establishments were derived from 2011 InfoUSA data. From listed grocery stores, two subsets were characterized—large grocery stores that typically sell fresh foods and convenience stores including local and national chains (e.g. Seven-Eleven). Fast food establishments were identified from listed fast food chain restaurants. In addition, any restaurant or convenience store whose name included fried chicken, sandwich, fries, burgers, hot dogs, shakes, pizza, drive through, and express was added to the unhealthy food outlet category. Further, all outlets listed as bars in the InfoUSA data base were identified. Finally, the outlet densities (i.e. outlets per census population) were characterized into three indices of food and beverage outlets: 1) healthy outlet density which included counts of large grocery stores as the numerator, 2) unhealthy outlet density which included the count of fast food outlets and convenience stores [66] as the numerator, and 3) bar density which included the count of all outlets listed as bars as the numerator.

- Physical Activity Environment Variables: As with food establishment data, commercially available data from InfoUSA were used to characterize the availability of physical activity conducive facilities using the Standard Industrial Classification (SIC) codes which correspond to those used by the US Census [53, 50]. A list of SIC codes representing physical activity-related facilities was compiled and used to identify and enumerate those facilities, including parks. The variable was characterized as the density of physical activity facilities.

- Walkability Variables: Walkability was defined as the degree of street connectivity in a census tract (i.e. neighborhood street networks that are continuous, integrated, and maximize linkages between starting points and destinations, providing multiple route options) [30, 67]. The indices of street connectivity calculated for the analysis are described in Table 1. They include intersection density, street density, and connected node ratio. Intersection density refers to the density of intersections in a neighborhood (i.e. more intersections per unit area, more connectivity). Street density refers to the degree to which a neighborhood has a high concentration of streets (i.e. more street miles per unit area, more connectivity). Finally, connected node ratio refers to the degree to which intersections in a neighborhood are of the types that increase connectivity (i.e. more four-way intersections yield more connectivity).

- Population Density was defined as total population per unit area (minus commercial, industrial and parkland) based on US Census estimates, with a higher percentage representing a higher population density.

- Crime Variable: The degree to which a neighborhood was exposed to crime was obtained from the ESRI. The ESRI data provides an index of violent crime at the census tract level. The index is generated by modeling city or county crime statistics to infer rates at the census tract level.

- Economic Deprivation Variables: Economic deprivation was measured using two different variables: 1) Income-to-poverty ratio and 2) concentrated disadvantage index. The income-to-poverty ratio is assessed at the individual level in the NHANES. It represents the ratio of family or unrelated individual income to their appropriate poverty threshold. Ratios below 1.00 indicate that the income for the respective family or unrelated individual is below the official definition of poverty, while a ratio of 1.00 or greater indicates income above the poverty level (U.S. Census Bureau, 2004). The

TABLE 7.2
The sample sizes of three population groups.

Population Group	Sample Size	Proportions
Non-Hispanic White	2994	57.14%
Hispanic White	1224	23.36%
Non-Hispanic Black	1022	19.50%

TABLE 7.3
The comparison of BMI and obesity among three population groups.

	NHW	HW	NHB	P-value
BMI (mean)	27.90	28.45	30.10	< .001
obesity (proportion %)	29.46	32.11	44.32	< .001

ratio was categorized into tertiles as low, medium, and high. An index of concentrated disadvantage [68] was generated at the census tract level. Concentrated disadvantage is derived from six census measures including the percent of families or households below the poverty line, percent of families receiving public assistance, percent of unemployed individuals in the civilian labor force, percent of population that is black, percent of population less than 18 years of age, and percent of families with children that have a female as the head of the household.

7.4.2 Disparities

The two outcomes considered are BMI (continuous, calculated as *weight*(in kilograms) divided by *height*($in\ meters)^2$) and obesity (binary as $BMI >$ 30 or not). In this study, We did not include races other than blacks and whites, multi-race heritages or Hispanic blacks since the sample sizes for these groups are very small in the Survey. The exposure variable is therefore a combination of the race and ethnicity of three groups: Non-Hispanic White (NHW), Hispanic White (HW), and Non-Hispanic Black (NHB). The sample size of each group is listed in Table 7.2. The comparisons among three groups in outcomes are shown in Table 7.3. In the table, the p-value for comparison in means of BMI was by the ANOVA F-test and in proportion of obesity was by Chi-square test. Both tests show that there are significant differences in obesity and BMI among the three racial and ethnic groups.

There are controversies in interpreting the "race" or "ethnic" effect. It is impossible to establish the "causal effect of race or ethnicity". However, the "effect of race/ethnicity" can be defined. [77] extensively discussed the challenges and different interpretations of race effect. We use their interpretation where "the effect of race involves the joint effects of race-associated physical phenotype (e.g. skin color), parental physical phenotype, genetic background, and cultural context when such variables are thought to be hypothetically manipulable and if adequate control for confounding were possible"[77]. Direct effect of race is interpreted as the remaining racial disparity if distributions of various risk factors across racial groups could be equalized. The indirect effect from a certain risk factor is the change in the health disparity if the distributions of the risk factor can be set as the same across racial groups, while distributions for other risk factors are kept as observed. With this interpretation, the hypothetical manipulation on race is not required. Instead, the interpretation was performed by framing around more manipulable risk factors, such as environmental and health care facility variables. For detailed discussion on explaining "race effect", the readers are referred to [77]. In the same vein, "ethnic effect" is interpreted.

7.4.3 Descriptive Analysis

In the analysis, we considered the racial and ethnic disparities together with multivariate outcomes (BMI and obesity). Usually a risk factor is considered as a potential third-variable (may significantly explain the racial/ethnic disparities) if two conditions are satisfied: first, the risk factor has to be significantly associated with at least one of the outcomes when other variables are adjusted; and second, the risk factor distributes differently among the population groups. Based on the joint data of the 2005–2006 NHANES and environmental factors, Table 7.4 lists test results for each potential risk factors for the two necessary conditions. The first two columns are the p-values of type-III tests for the corresponding risk factors when all variables are used as predictors in a (generalized) linear model for BMI and obesity separately. Using NHW as the reference group, the last two columns list the p-values for testing whether the corresponding risk factors distributed differently when compared with the NHB and the HW group, respectively. Since the tests are based on linear correlations only and to avoid missing any significant third-variables, we set the significance levels at 0.1, higher than 0.05. All factors that satisfy the two necessary conditions are treated as potential third-variables. If a factor satisfies the first condition but not the second one, it is treated as a covariate. If a factor does not satisfy the first condition, it is left out for further analysis. However, if a factor is empirically considered as important, it can be forced into analysis as a potential third-variable even if the first condition is not satisfied. As a result of tests, eleven variables were selected as potential third-variables (marked by * in the table). The variable "male" was selected as a covariate. In addition, the variables "Street Density", "Connected Node Ratio" and "Intersection Density", which measure the walkability of living environment, are forced in as potential third-variables. It is informative to make an inference on their joint effect in explaining the racial/ethnic disparities.

TABLE 7.4
Test results on selecting potential third-variables.

	P1(bmi)	P1(obesity)	P2(NHW vs NHB)	P2(NHW vs HW)
Elevation *	0.000	0.000	0.000	0.000
Population Density	0.209	0.680	0.000	0.000
CDI *	0.034	0.004	0.000	0.000
Physical Activity *	0.000	0.000	0.299	0.005
VCI *	0.342	0.098	0.000	0.000
Male	0.022	0.000	0.540	0.417
Poverty	0.919	0.998	0.022	0.000
Foreign Born *	0.000	0.000	0.268	0.000
Smoker *	0.000	0.000	0.053	0.222
EnergyIntake *	0.087	0.169	0.053	0.468
age *	0.015	0.001	0.000	0.000
Street Density -	0.653	0.918	0.000	0.000
Unhealthy Outlet *	0.602	0.050	0.000	0.202
Connected Node Ratio -	0.457	0.430	0.000	0.000
BarDensity *	0.176	0.024	0.000	0.000
Intersection Density -	0.440	0.557	0.000	0.000
Convenient	0.841	0.318	0.000	0.015
Sugar Beverage *	0.095	0.250	0.000	0.000
Black	0.000	0.000	NA	NA
Hispanic	0.000	0.000	NA	NA

P1 is the p-value of type-III test for the risk factor in a (generalized) linear model for
BMI and obesity separately. P2 is the p-value for testing whether the risk factor distributed
differently between two population groups. *:variable selected as a potential third-variable;
-:variable forced in as a potential third-variable.

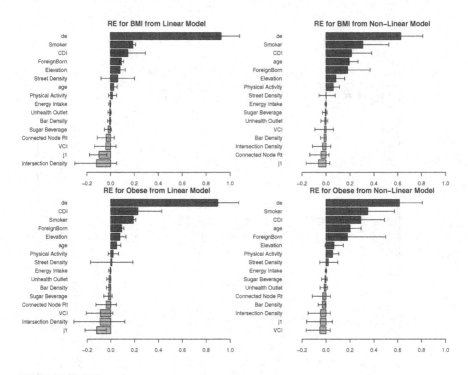

FIGURE 7.2
Relative Effects of Risk Factors that Explain Racial Disparities in BMI (upper) and obesity (lower) using Linear (left) and Non-Linear (right) Models. "j1" is the joint effect of the three walkability measurements.

7.4.4 Results on Racial Disparities

We used both linear and nonlinear models to build the relationships among variables. The relative effects, defined as the direct/indirect effect divided by the total effect, on each outcome are shown in Figure 7.2, with the left column the results from generalized linear models and right column from nonlinear models. The confidence bounds of relative effects are also presented. The upper row is for the outcome BMI and the lower row for obesity.

The inferences on indirect effects of risk factors and the joint effect of walkability (measured by "street density", "connected node ratio" and "intersection density") are summarized in Table 7.5. The 95% confidence interval (in parenthesis) for each indirect effect is calculated from the confidence ball of the bootstrap results centered at the estimates based on the whole data set.

We found from the summary that using the nonlinear models, risk factors can partially explain the disparities in BMI (37.3%) and obesity (38.5%). The direct effects of race became significantly less (the 95% confidence intervals

TABLE 7.5
Inferences on indirect effects of risk factors in explaining racial disparities in outcomes.

	Linear Model		Non-Linear Model	
	bmi	obesity	bmi	obesity
Elevation	0.17(0.07,0.28)	0.05(0.00,0.10)	0.17(0.07,0.28)	0.05(0.00,0.10)
CDI	0.32(-.18,0.82)	0.15(-.02,0.32)	0.25(-.09,0.87)	0.09(-.00,0.29)
Physical Activity	0.01(-.08,0.16)	0.00(-.03,0.05)	0.08(-.04,0.27)	0.02(-.01,0.08)
VCI	-.09(-.38,0.20)	-.06(-.17,0.08)	-.01(-.23,0.22)	-.01(-.11,0.08)
Foreign Born	0.21(0.13,0.29)	0.07(0.04,0.09)	0.40(0.09,1.09)	0.08(-.04,0.39)
Smoker	0.41(0.26,0.61)	0.13(0.08,0.19)	0.45(0.13,1.36)	0.23(0.06,0.45)
Energy Intake	-.01(-.05,0.01)	0.00(-.02,0.01)	0.00(-.03,0.01)	0.00(-.01,0.01)
age	0.07(-.02,0.15)	0.04(0.01,0.06)	0.33(0.19,0.54)	0.10(0.06,0.16)
Street Density	0.14(-.32,0.58)	0.01(-.19,0.17)	-.01(-.20,0.21)	-.01(-.08,0.09)
Unhealthy Outlet	-.02(-.07,0.03)	-.10(-.03,0.00)	-.00(-.13,0.05)	-.01(-.03,0.01)
Connected Node Ratio	-.08(-.35,0.18)	-.03(-.12,0.07)	-.04(-.31,0.20)	-.00(-.08,0.05)
Bar Density	-.02(-.07,0.03)	-.01(-.04,0.00)	-.01(-.10,0.01)	-.01(-.04,0.00)
Intersection Density	-.27(-.85,0.23)	-.06(-.27,0.12)	-.03(-.32,0.20)	-.02(-.11,0.06)
Sugar Beverage	-.05(-.15,0.05)	-.01(-.05,0.02)	-.01(-.08,0.05)	0.00(-.03,0.02)
Walkability (joint)	-.22(-.49,-.01)	-.08(-.18,0.01)	-.08(-.40,0.14)	-.03(-.12,0.08)
Race (direct effect)	2.03(0.18,2.88)	0.60(0.31,0.89)	1.17(0.20,2.10)	0.36(0.08,0.69)
Total Effect	2.18(1.58,2.88)	0.60(0.47,0.89)	1.72(1.07,2.76)	0.51(0.28,0.83)

of relative effects are to the left of 1) when other factors are adjusted. For generalized linear models, less disparities were explained (BMI 7.4% and obesity 10.8%). Also the direct effect of race did not change significantly after adjusting for other variables although some variables have significant indirect effect in explaining the racial disparities. We found that for both linear and nonlinear models, "elevation", "foreign born", "smoker", "cdi" and "age" can partially explain the racial disparities in obesity and/or bmi. "physical activity" is shown to have an influence on the racial disparities by nonlinear models, while the joint effect of walkability measurements is significant in explaining obesity disparity by linear model.

7.4.5   Results on Ethnic Disparities

Table 7.6 lists the third-variable effects of risk factors in explaining the ethnic disparities in BMI and obesity separately.

We also drew figures to describe the third-variable effects of all potential third-variables using the *plot* function provided by the *mma* package. The graphs are provided as online supplementary files at the book website.

TABLE 7.6
Inferences on indirect effects of risk factors in explaining ethnic disparities in outcomes.

	Linear Model		Non-Linear Model	
	bmi	obesity	bmi	obesity
Elevation	-.07(-.02,-.15)	-.02(-.05,0.00)	-.02(-.16,0.05)	-.01(-.05,0.04)
CDI	0.14(-.08,0.38)	0.07(-.01,0.14)	0.20(0.02,0.52)	0.08(0.03,0.22)
Physical Activity	-.08(-.23,0.02)	-.03(-.07,0.01)	-.02(-.17,0.09)	-.00(-.06,0.03)
VCI	-.04(-.14,0.09)	-.02(-.08,0.04)	-.01(-.21,0.11)	-.01(-.09,0.05)
Foreign Born	-.49(-.79,-.17)	-.16(-.29,-.03)	-.09(-.75,-.01)	-.06(-.20,-.04)
Smoker	0.13(0.01,0.26)	0.03(-.02,0.07)	-.02(-.38,-.24)	0.01(-.11,0.10)
Energy Intake	-.00(-.03,0.02)	-.00(-.01,0.01)	0.00(-.02,0.01)	0.00(-.00,0.00)
age	0.08(-.02,0.20)	0.04(0.00,0.08)	0.21(0.09,0.04)	0.07(0.04,0.14)
Street Density	0.12(-.31,0.54)	0.01(-.12,0.14)	-.02(-.12,0.19)	-.01(-.05,0.07)
Unhealthy Outlet	-.01(-.01,0.03)	0.00(-.01,0.16)	0.01(-.03,0.08)	0.01(-.02,0.03)
Connected Node Ratio	-.09(-.36,0.23)	-.03(-.12,0.08)	0.00(-.19,0.21)	0.00(-.09,0.05)
Bar Density	-.02(-.07,0.02)	-.01(-.03,0.00)	-.01(-.09,0.02)	-.01(-.04,0.00)
Intersection Density	-.19(-.58,0.17)	-.04(-.18,0.09)	-.02(-.25,0.12)	-.04(-.11,0.03)
Sugar Beverage	-.05(-.16,0.05)	-.01(-.05,0.03)	-.01(-.08,0.04)	0.00(-.03,0.01)
Walkability (joint)	-.16(-.37,-.01)	-.06(-.14,0.01)	-.04(-.22,0.14)	-.04(-.13,0.03)
Race (direct effect)	1.56(0.87,2.23)	0.41(0.12,0.69)	-.08(-.05,0.71)	0.36(-.03,0.20)
Total Effect	0.57(0.01,1.01)	0.12(-.11,0.30)	0.06(-.41,0.66)	0.01(-.16,0.20)

8

Regularized Third-Variable Effect Analysis for High-Dimensional Dataset

In data analysis, we often have to deal with high-dimensional datasets, where the number of variables is very large. Sometimes, the number of variables can be larger than the number of observations. This can also happen in the third-variable effect analysis, where there are high-dimensional third variables. If the non-linear method is chosen with the proposed general third-variable analysis and the R package *mma* is used, the high-dimensional problem is handled automatically since the tree-based method is designed to deal with high-dimensional data. However, if the linear method is preferred, high-dimensional data can bring in problems. For example, if the number of variables is larger than the number of observations, a linear model cannot provide converged estimates for all explanatory variables. In such cases, we propose to use the elastic net regularized linear regression in multiple third-variable analysis. In exploring the exposure-third variable-outcome relationship, we put a penalty function on coefficients of explanatory variables in predicting the outcome. The penalization on the coefficient is inversely proportional to the association between the exposure variable and each third-variable. Therefore, in estimating the effect of a third variable, the exposure-third variable and the third variable-outcome associations are jointly considered. An R package, *mmabig*, is compiled for the proposed method. We perform a series of sensitivity and specificity analyses to examine factors that can influence the power of identifying important third-variables. Further, we illustrate how to consider potential nonlinear associations among variables in the high-dimensional third-variable analysis. Simulation studies have shown that the proposed method consistently obtains larger power when compared with its main competitors. Further, the method is used with a real data set to explore factors that contribute to the racial disparity in survival rates among breast-cancer patients. This chapter is derived in part from an article published in the Journal Statistics and Its Interface [47] on July 8, 2021, available at https://dx.doi.org/10.4310/21-SII664.

DOI: 10.1201/9780429346941-8

8.1 Regularized Third-Variable Analysis in Linear Regression Setting

The third-variable analysis based on linear regressions uses linear models to build the relationship among variables. For the simple scenario, there are one predictor X, one outcome Y and one third variable M. When all the variables are continuous, the relationships among variables can be modeled as follows:

$$
\begin{aligned}
m_i &= \alpha_0 + \alpha_1 x_i + \epsilon_{i1}; \\
y_i &= \beta_0 + \beta_1 m_i + \beta_2 x_i + \epsilon_{i2}; \\
\epsilon_{ij} &\overset{ind}{\sim} N(0, \sigma_j^2), \quad j = 1, 2; \ i = 1, \ldots, n.
\end{aligned}
$$

In such setting, the indirect effect of M, is typically measured by $\alpha_1 \beta_1$, the product of the coefficient of X when it is regressed on M, and the coefficient of M in explaining Y controlling for X [49]. To explain the indirect effect, when X changes by a unit, M changes by α_1 unit, and when M changes by one unit, Y changes by β_1 unit. Therefore, a one-unit change in X results in $\alpha_1 \times \beta_1$ units change in Y through M. This is the product-of-coefficients method ("CP") that has been discussed in Chapter 2. Also, the result is consistent with the general third-variable analysis method discussed in the previous chapters.

When there are multiple third-variables, a conceptual mediation model is presented in Figure 8.1. We assume that there are one exposure variable, denoted as $\mathbf{x} = \{x_1, \ldots, x_n\}$, where n is the number of observations; p third-variables: $\mathbf{m} = \{\mathbf{m}_1, \ldots, \mathbf{m}_p\}$, where $\mathbf{m}_j^T = (m_{j1}, \ldots, m_{jn})$ and $j = 1, \ldots, p$; and one outcome, $\mathbf{y} = \{y_1, \ldots, y_n\}$. In Figure 8.1, the lines (with no arrows) among third-variables indicate that given the exposure variable, third-variables can be associated with each other. Other covariates, $\mathbf{z}$, are variables that are significantly related with the outcome but not with the exposure variable. For simplicity, we ignore $\mathbf{z}$ for now. As discussed in Chapter 7, the exposure variable and the outcome can be extended to be multivariate, but our discussion in this section focuses on one exposure and one outcome only.

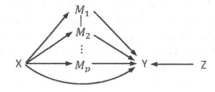

FIGURE 8.1
The conceptual model of multiple third-variable analysis.

In the linear-model setting, to make inferences on third-variable effects, $p + 1$ linear regressions are needed. The first model is on $\mathbf{y}$ given $\mathbf{x}$ and $\mathbf{m}$. The linear regression has the following format:

$$y_i = \beta_0 + \beta_x x_i + \beta_1 m_{1i} + \ldots + \beta_p m_{pi} + \epsilon_{0i}, \quad \epsilon_{0i} \overset{iid}{\sim} N(0, \sigma_0^2). \quad (8.1)$$

In addition, there are p linear regressions, each modeling the relationship between a third-variable, $\mathbf{m}_j$, and the exposure variable such that

$$m_{ji} = \alpha_{0j} + \alpha_j x_i + \epsilon_{ji}, \quad j = 1, \ldots, p; \quad (8.2)$$

where $\epsilon_i^T = (\epsilon_{1i}, \ldots, \epsilon_{pi})$ has an independent multivariate normal distribution with the mean vector $\mathbf{0}_p$ and variance-covariance matrix Σ, for $i = 1, \ldots, n$. Note that Σ does not have to be a diagonal matrix, indicating that given x, third-variables are allowed to be associated with each other.

Under the above setting, the general multiple third-variable analysis proposed in previous chapters generates the same third-variable effect estimates as the CP method where the direct effect from the exposure variable is β_x, the indirect effect from the jth third-variable is $\alpha_j \beta_j$, and the total effect is $\beta_x + \sum_{j=1}^p \alpha_j \beta_j$ (generalization of Theorem 4.2 in Chapter 4).

The purpose of third-variable effect analysis is to identify third-variables that have significant indirect effects and estimate those effects. When there is a large number of potential third-variables, we propose to estimate the coefficients of regression model (8.1), and therefore the third-variable effects, by minimizing the penalized function:

$$L(\lambda, \gamma, \hat{\alpha}, \beta) = \sum_{i=1}^n \left(y_i - \beta_0 - \beta_x x_i - \sum_{j=1}^p \beta_j m_{ji} \right)^2 + \quad (8.3)$$

$$\frac{\lambda(1-\gamma)}{2} \left[\sum_{j=1}^p (\hat{\alpha}_j \beta_j)^2 + \beta_x^2 \right] + \lambda\gamma \left[\sum_{j=1}^p |\hat{\alpha}_j \beta_j| + |\beta_x| \right],$$

for some $\lambda \geq 0$ and $\gamma \in [0, 1]$, where $\hat{\alpha}s$ are the coefficient estimates for models (8.2). The penalization function is the same as that for the elastic-net regression, except that the penalty is on third-variable effects instead of just coefficients. Inheriting the good properties of the elastic net regression (see Chapter 3), the proposed method can identify significant third-variable effects, where indirect effects, $\alpha\beta$, and/or the direct effect, β_x, are significantly different from 0, and estimate these third-variable effects at the same time.

8.2 Computation: The Algorithm to Estimate Third-variable Effects with Generalized Linear Models

There are established computational methods to estimate the coefficients from the elastic net regularized generalized linear models. The *glmnet* package implements one of the computational methods and is used for the proposed third-variable analysis. *glmnet* can deal with Gaussian, binomial, Poisson, multinomial and time-to-event (using cox model) types of outcomes. Readers are referred to Chapter 3 for details about using the package.

We propose an algorithm to use the elastic net computational method to make inferences on third-variable effects of interests. In the algorithm, we also allow the use of any predictive models to fit relationships between the exposure variable and each third-variable, therefore enable the fitting of potential nonlinear relationships and allow the third-variables to be of different types. Model (8.2) is generalized to have the following format

$$E(m_{ji}|x_i) = l_j^{-1}\left(g_j(x_i)\right), \quad j = 1,\dots,p; \tag{8.4}$$

where l_j is the link function that links the mean of M_j with the prediction function g_j. g_j is any predictive model that predicts m_j using x. For example, to use the generalized spline models to fit the relationship between the exposure variable and third-variables, model (8.4) is

$$E(m_{ji}|x_i) = l_j^{-1}\left(\alpha_{0j} + \sum_{k=1}^{K}\alpha_{kj}h_k(x_i)\right), \quad j = 1,\dots,p;$$

where h_k is the spline basis function, and K basis functions are used to fit m_j with x. Using the general predictive models, Algorithm 8 presents the procedure of estimating third-variable effects.

Algorithm 8 *Algorithm to estimate third-variable effects.*

1. For each third-variable, m_j, fit the general model (8.4) that predicts m_j using x.

2. Based on the models fitted from step 1, for $j = 1,\dots,p$:

•If x is continuous, calculate the average changing rate in m_j when x changes by a margin mg,

$$\Delta M_j = \frac{1}{n}\sum_{i=1}^{n}\frac{E(m_{ji}|x_i = x_i + mg) - E(m_{ji}|x_i = x_i)}{mg}.$$

•If x is binary, calculate the change in the mean of m_j when x changes from 1 to 0,

$$\Delta M_j = E(m_{ji}|x_i = 1) - E(m_{ji}|x_i = 0).$$

3. Transform all third-variables such that $m_{ij}^* = m_{ij}/\Delta M_j$, for $i = 1,\dots,n$ and $j = 1,\dots,p$.

4. Fit a linear regression model where y_i is the response variable, x_i and $m_{ij}^*, j = 1,\dots,p$ are covariates, using the elastic net penalized function to estimate coefficients.

  5. *The average effect of each variable is its estimated coefficient from elastic net regression: direct effect for x and indirect effect for M_j.*

For the regularized third-variable analysis and its computational algorithm, we have the following comments:

- ΔM_j measures the average changing rate in M_j when x changes by a margin, mg. In general, we set $mg = 1$. However, the margin can be changed.

- If the linear regression model (8.2) is used to fit the relationship between x and m_j, α_j is the ΔM_j. In addition, potential nonlinear relationship between third-variables and the exposure variable can be fitted through different predictive models. As is shown in the algorithm, ΔM_j can be calculated with any predictive models.

- For continuous third-variables, ΔM_j denotes the average changing rate over x. The changing rate can be different at different value of x. Therefore, the indirect effect of M_j can alter with x. Algorithm 8 gives the process of estimating the average effect. Based on the results, the indirect effect of M_j at different x is also estimable. Denote the average indirect effect of M_j as AIE_j, the indirect effect of M_j at $X = x$ is $\frac{AIE_j}{\Delta M_j} \times \frac{E(m_j|X=x+mg)-E(m_{ji}|X=x)}{mg}$.

- In step 4, a linear regression model is fitted with elastic net method. However, elastic net regularized regression has already been extended to any generalized linear models to deal with any types (e.g. categorical, continuous, time-to-event) of response variable(s). For example, the *glmnet* package can deal with Gaussian, Binomial, Poisson, Multinomial, and time-to-event (using cox model) type of outcomes. Therefore, generalized linear models can also be used in step 4 to deal with different types of the response variable for the third-variable analysis. Note that when different model than 8.1 is used, the explanation of third-variable effect needs to be modified accordingly. For example, if a logistic regression is used for binary outcome, the direct and indirect effect is calculated in terms of $logit(y = 1)$ but not of the probability of $y = 1$. Transformation should be made to explain the effect in terms of the probability of $y = 1$.

- For elastic net regression, covariates need to be standardized to have the same mean and standard deviation, so that all coefficients are shrunk at the same scale. For third-variable effect estimation, the standardization of covariates is not required since third-variable effects are calculated at the same scale: the changing rate in Y when X changes at the margin mg. Hence the estimated third-variable effects are scale-invariant.

- Although we discussed only the binary and continuous exposures in the algorithm, the method is readily extended to multivariate exposure variables (See Chapter 7). As an example, a multi-categorical exposure variable can be handled by transforming it to multiple binary exposures.

- Bootstrap method can be used with the algorithm to estimate uncertainties.

8.3   The R Package: mmabig

The *mmabig* package [81], available on the Comprehensive R Archive Network (CRAN), was generated based on the above algorithm for third-variable effect inferences with high-dimensional datasets. In the package, *glmnet* by [70] is used for the elastic net estimates. To use the R package *mmabig*, we need to first install the package in R and load it.

```
install.packages("mmabig")
library(mmabig)
```

8.3.1   Simulate a Dataset

To illustrate how to use the *mmabig* package, we simulated a dataset. The following code generates a simulated data set with 20 potential third-variables, of which five (*m16-m20*) are multicategorical variables. The real mediator/confounders are *m11, m12, and m16*, which highly relate with both the predictor *pred* and the outcome *y*.

```
# a binary predictor
set.seed(1)
n=100
pred=rbinom(n,1,0.5)
m1=matrix(rnorm(n*10),n,10)
m2<-matrix(rnorm(n*10,mean=pred,sd=1),n,10)
m3.1=m2[,6:10]
m3=m3.1
m2=m2[,1:5]
m3[m3.1<=0.1]=0
m3[0.1<m3.1 & m3.1<=1]=1
m3[m3.1>1]=2
m3<-apply(m3,2,as.factor)
m<-data.frame(m1,m2,m3)
colnames(m)<-c(paste("m0",1:9,sep=""),
               paste("m",10:20,sep=""))

lu<--0.5363+0.701*pred+0.801*m[,1]+0.518*m[,2]+1.402*m[,11]+
    0.773*m[,12]+ifelse(m[,16]=="2",2.15,0)+
```

```
ifelse(m[,16]=="1",0.201,0)
```

```
# a continuous y
y<-rnorm(n,lu,1)
```

Summarize the codes, 20 third-variables and one outcome are generated in the following way:

$$m_{ij} \sim N(0,1), j = 1, \ldots, 10;$$
$$m_{ij} \sim N(x_i, 1), j = 11, \ldots, 15;$$
$$m_{ij.1} \sim N(x_i, 1),$$
$$m_{ij} = \begin{cases} 1 \text{ if } m_{ij.1} > 1, \\ 0 \text{ if } m_{ij.1} \leq 0.1, \\ 2 \text{ otherwise, } j = 16, \ldots, 20; \end{cases}$$
$$y_i = 0.5363 + 0.701 \times x_i + 0.801m_{i1} + 0.518m_{i2} + 1.402m_{i,11}$$
$$+ 0.773m_{i,12} + 2.15I(m_{i,16} = 2) + 0.201I(m_{i,16} = 1) + \epsilon_i;$$

where x_i is the exposure variable, $i = 1, \ldots, 100$, and $\epsilon_i \sim N(0,1)$ is the random error. By the data generation mechanism, m_{15} to m_{20} are categorical variables with three levels; m_{11}, m_{12} and m_{16} are mediators; and m_1 and m_2 are covariates. All other variables, related with the exposure or not, are not relevant variables in the third-variable analysis.

8.3.2 Function *data.org.big*

The package *mmabig* provides a step-by-step process for the third-variable effect analysis. The function *data.org.big* identifies potential third-variables and covariates through multiple tests. As described in previous chapters, a potential third-variable is identified if it is significantly associated with both the exposure and the outcome variables. In this function, two methods can be chosen to test whether the third-variable is significantly associated with the outcome: univariate test or elastic net regression that jointly considers all variables together. A covariate is identified if it is significantly related with the outcome but not with the exposure variable. The function *data.org.big* returns an object of "med_iden" class that organizes the data into a format to be used directly for the mediation analysis functions.

8.3.2.1 Univariate Exposure and Univariate Outcome

Specifically in the function, The exposure variable is specified by "pred=". All potential third-variables and covariates are included in the data frame "x". The outcome variable is specified by "y". Both the exposure and the outcome can be multivariate. The argument "mediator" indicates the variables in "x" that should be tested for third-variables or covariates. Those variables are listed in "mediator" as column numbers or variable names in "x". For example, the argument "mediator =1:ncol(m)" indicates that all variables in x should be tested as potential third-variables or covariates. In addition to the "mediator" argument, those variables can also be identified by "contmed", "binmed" and "catmed", where binary, categorical and continuous potential third-variables in "x" are identified by "binmed", "catmed" and "contmed", respectively. If "mediator" is used instead, binary and categorical variables are identified if they are factors or characters and their reference group is set to be the first level of the factor or factorized character. If a variable in "mediator" has only two unique values, the variable will also be identified as binary. For a binary or categorical variable, if the reference group needs to be changed rather than using the first level of the variable, the binary or categorical variable should be listed in "binmed" or "catmed", and the corresponding reference group in "binref" or "catref" in order.

Two tests are performed to identify a potential third-variable: first, to check if the variable is significantly related with the exposure variable adjusting for other covariates. The significance level is set by the argument "alpha2", whose default value is 0.01. Second, to check if the variable is significantly related with the outcome not adjusting (by setting "testtype=2", by default) or adjusting ("testtype=1") for the exposure(s) and other variables. The significance level is set by "alpha1". A variable that is significant in the second test but not the first test is included for further analysis as a covariate. Variables that are not significant from the second test are removed for further analysis.

Variables in "x" but not assigned in "mediator", "contmed", "binmed" and/or "catmed" are specified and therefore forced in further analysis as covariates without being tested. Variables can also be forced in analysis as third-variables by being included in the argument "jointm". The use of "jointm" is the same as it is in the *mma* package.

The argument "alpha" is the elasticnet mixing parameter such that $0 \leq alpha \leq 1$, with "alpha=1" be the lasso penalty, and "alpha=0" be the ridge penalty. By default, "alpha=1". Another argument for the elasticnet regularization method is "lambda", which lists the sequence used for cross-validation with *cv.glmnet*. By default, it is set as $exp(seq(log(0.001), log(5), length.out = 15)$. Finally, the weight for each observation can be set by the argument "w". By default, all observations are equally weighted. The following is a sample code of performing the data organization and tests using the simulated data.

```
data.e1<-data.org.big(x=m,y=data.frame(y),mediator=
        1:ncol(m), pred=data.frame(pred),testtype=1)
summary(data.e1,only=TRUE)
```

The results from *data.org.big* function are summarized by the *summary* function. When "only=TRUE", the summary only shows the test results for selected covariates and third-variables. Results from the above codes are shown as follows:

```
Identified as mediators:
```

```
[1] "m11"   "m12"   "m161"  "m162"
Selected as covariates:
[1] "m01"  "m02"
Tests:
      Coefficients.y P-Value 2.pred
m01             0.702          0.548
m02             0.534          0.920
m11 *           1.180          0.000
m12 *           0.479          0.000
m161 *          0.000          0.000
m162 *          1.548          0.000
----

*:mediator,-:joint mediator
Coefficients: estimated coefficients;
P-Value 2:Tests of relationship with the Predictor
```

For the tests results, the first column shows the results for the second test. Since "testtype" is 1 in the code, elasticnet was used to fit the full model. The coefficients from the full model is listed. A coefficient of 0 indicates that the coefficient is shrunk to 0, i.e. the variable is not selected in the full model. If "testtype=2", univariate test is performed to test if each variable is related with the outcome. When testtype=2, the first column is labeled as "P-Value 1", which gives the p-value for the test of each variable. The second column is labeled as "P-Value 2", which are the results for the first test. In the above result, the multicategorical third-variable $m16$ has three categories. It has been binarized into two binary variables $m161$ and $m162$, both of which are selected together as potential third-variables or covariates or removed together for further analysis. The following codes and results are generated when "testtype=2". For the simulated data, both methods selected the same third-variables and covariates correctly.

The R codes:

```
data.e1.2<-data.org.big(x=m,y=data.frame(y),mediator=1:ncol(m),
                        pred=data.frame(pred))
summary(data.e1.2,only=TRUE)
```

Output from R:

```
Identified as mediators:
[1] "m11"   "m12"   "m161"  "m162"
Selected as covariates:
[1] "m01"  "m02"
Tests:
      P-Value 1.y P-Value 2.pred
m01         0.001          0.548
m02         0.000          0.920
m11 *       0.000          0.000
m12 *       0.000          0.000
```

```
m16 *        0.000              0.000
----
```

*:mediator,-:joint mediator
P-Value 1:univariate relationship test with the outcome;
P-Value 2:Tests of relationship with the Predictor

8.3.2.2 Survival Outcome

The functions in mmabig can deal with binary, categorical, continuous, or time-to-event outcomes. If the outcome is time-to-event, it should be defined by the *Surv* function in the R package *survival*. In such case, the outcome y should be a two-column matrix with columns of time and event status. A cox-hazard model will be fitted as the full model. The following is an example with simulated time-to-event data and the results of the *data.org.big* function. **The R codes:**

```
lambda=1/500
survt=-log(runif(n))/lambda/exp(lu)
st=round(runif(n,1,500),0)
time=ifelse(st+survt>600,600,st+survt)-st
cen=ifelse(st+survt>600,0,1)
y=Surv(time,cen)

data.e3<-data.org.big(x=m,y=data.frame(y),mediator=1:ncol(m),
                      pred=data.frame(pred),testtype=1)
summary(data.e3,only=TRUE)
```

Output from R:

```
Identified as mediators:
[1] "m11"  "m12"  "m161" "m162"
Selected as covariates:
[1] "m01" "m02" "m05"
Tests:
        Coefficients.y P-Value 2.pred
m01             0.319          0.548
m02             0.062          0.920
m05             0.020          0.485
m11 *           0.838          0.000
m12 *           0.408          0.000
m161 *          0.000          0.000
m162 *          1.211          0.000
----
```

*:mediator,-:joint mediator
Coefficients: estimated coefficients;
P-Value 2:Tests of relationship with the Predictor

8.3.2.3   Multivariate Predictors and/or Outcomes

In addition, the package can handle multivariate and multicategorical predictors/exposures. If the predictor is multicategorical of k levels, *data.org.big* first transforms the predictor to $k-1$ binary predictors. If a variable significantly relates with any of the $k-1$ predictors, the variable passes the first test described above. P-value 2 is shown for each predictor. In the following example, the predictor has three levels.

The R codes:

```
# multicategorical predictor
set.seed(1)
n=100
pred=rmultinom(100,1,c(0.5, 0.3, 0.2))
pred=pred[1,]*0+pred[2,]*1+pred[3,]*2
m1=matrix(rnorm(n*10),n,10)
m2<-matrix(rnorm(n*10,mean=pred,sd=1),n,10)
m3.1=m2[,6:10]
m2=m2[,1:5]
m3=m3.1
m3[m3.1<=0.1]=0
m3[0.1<m3.1 & m3.1<=1]=1
m3[m3.1>1]=2
m3<-apply(m3,2,as.factor)
m<-data.frame(m1,m2,m3)
colnames(m)<-c(paste("m0",1:9,sep=""),paste("m",10:20,sep=""))
pred<-as.factor(pred)
# continuous y
lu<--0.5363+ifelse(pred=="1",0.3,0)+ifelse(pred=="2",0.7,0)+
    0.801*m[,1]+0.518*m[,2]+1.402*m[,11]+0.773*m[,12]+
ifelse(m[,16]=="2",2.15,0)+ifelse(m[,16]=="1",0.201,0)
y<-rnorm(n,lu,1)

data.m.e1<-data.org.big(x=m,y=data.frame(y),mediator=1:ncol(m),
                   pred=data.frame(pred),testtype=1)
summary(data.m.e1,only=TRUE)
```

Output from R:

```
Identified as mediators:
[1] "m11" "m12" "m15"
Selected as covariates:
[1] "m01" "m02" "m16"
Tests:
```

	Coefficients.y	P-Value 2.pred1.i.1	P-Value 2.pred1.i.2
m01	0.546	0.721	0.732
m02	0.461	0.157	0.201
m11 *	1.375	0.000	0.000
m12 *	0.609	0.000	0.000
m15 *	0.041	0.000	0.000

```
----
*:mediator,-:joint mediator
Coefficients: estimated coefficients;
P-Value 2:Tests of relationship with the Predictor
```

In the following example, the predictor is bivariate and both are continuous.

The R codes:

```
# multivariate predictor
set.seed(1)
n=100
pred=cbind(runif(n,-1,1),rnorm(n))
m1=matrix(rnorm(n*10),n,10)
m2<-matrix(rnorm(n*5,mean=0.3*pred[,1]+0.4*pred[,2],sd=1),n,5)
m3.1=matrix(rnorm(n*5,mean=0.7*pred[,1]+0.8*pred[,2],sd=1),n,5)
m3=m3.1
m3[m3.1<=0]=0
m3[0<m3.1 & m3.1<=1.28]=1
m3[m3.1>1.28]=2
m3<-apply(m3,2,as.factor)
m<-data.frame(m1,m2,m3)
colnames(m)<-c(paste("m0",1:9,sep=""),paste("m",10:20,sep=""))
colnames(pred)=c("x1","x2")
# binary y
lu<--0.6852+0.3*pred[,1]+0.7*pred[,2]+0.801*m[,1]+0.518*m[,2]+
    1.402*m[,11]+0.773*m[,12]+ifelse(m[,16]=="2",2.15,0)+
 ifelse(m[,16]=="1",0.201,0)
y<-rbinom(n,1,exp(lu)/(1+exp(lu)))

data.m.c.e2<-data.org.big(x=m,y=data.frame(y),mediator=1:ncol(m),
 pred=data.frame(pred),testtype=1,alpha1=0.05,alpha2=0.05)
summary(data.m.c.e2,only=TRUE)
```

Output from R:

```
Identified as mediators:
[1] "m11"  "m12"  "m15"  "m161" "m162"
Selected as covariates:
[1] "m01" "m06"
Tests:
```

	Coefficients.y	P-Value 2.x1	P-Value 2.x2
m01	0.187	0.607	0.533
m06	-0.017	0.216	0.977
m11 *	0.436	0.653	0.000
m12 *	0.225	0.003	0.000
m15 *	0.080	0.095	0.000
m161 *	0.000	0.003	0.003
m162 *	0.365	0.000	0.000

```
----
*:mediator,-:joint mediator
Coefficients: estimated coefficients;
P-Value 2:Tests of relationship with the Predictor
```

Furthermore, the package can deal with multivariate outcomes. The following codes generate both multivariate predictors and multivariate responses. The results from *data.org.big* are summarized for each combination of the exposure-outcome relationship.

The R codes:

```
# multivariate predictor
set.seed(1)
n=100
pred=cbind(runif(n,-1,1),rnorm(n))
m1=matrix(rnorm(n*10),n,10)
m2<-matrix(rnorm(n*5,mean=0.3*pred[,1]+0.4*pred[,2],sd=1),n,5)
m3.1=matrix(rnorm(n*5,mean=0.7*pred[,1]+0.8*pred[,2],sd=1),n,5)
m3=m3.1
m3[m3.1<=0]=0
m3[0<m3.1 & m3.1<=1.28]=1
m3[m3.1>1.28]=2
m3<-apply(m3,2,as.factor)
m<-data.frame(m1,m2,m3)
colnames(m)<-c(paste("m0",1:9,sep=""),paste("m",10:20,sep=""))
colnames(pred)=c("x1","x2")
#multivariate responses
y<-cbind(rnorm(n,lu,1),rbinom(n,1,exp(lu)/(1+exp(lu))))
colnames(y)=c("y1","y2")

data.m.m.c.e2<-data.org.big(x=m,y=data.frame(y),mediator=1:ncol(m),
pred=data.frame(pred), testtype=1,alpha1=0.05,alpha2=0.05)
summary(data.m.m.c.e2,only=TRUE)

data.m.m.c.e2.2<-data.org.big(x=m,y=data.frame(y),mediator=1:ncol(m),
pred=data.frame(pred),alpha1=0.05,alpha2=0.05)
summary(data.m.m.c.e2.2,only=TRUE)
```

Output from R:

```
> summary(data.m.m.c.e2,only=TRUE)
Identified as mediators:
 [1] "m11"  "m12"  "m15"  "m161" "m162" "m171" "m172" "m191" "m192"
     "m201" "m202"
Selected as covariates:
[1] "m01" "m02" "m09"
Tests:
      Coefficients.y1 Coefficients.y2 P-Value 2.x1 P-Value 2.x2
m01             0.652           0.000        0.607        0.533
m02             0.454           0.000        0.053        0.936
```

m09	0.101	0.000	0.678	0.292
m11 *	1.375	0.541	0.653	0.000
m12 *	0.661	0.000	0.003	0.000
m15 *	0.059	0.000	0.095	0.000
m161 *	0.547	0.000	0.003	0.003
m162 *	1.519	0.000	0.000	0.000
m171 *	0.000	0.000	0.219	0.219
m172 *	0.163	0.000	0.000	0.000
m191 *	0.000	0.000	0.028	0.028
m192 *	0.072	0.000	0.000	0.000
m201 *	0.000	0.000	0.053	0.053
m202 *	0.149	0.000	0.000	0.000

```
----
*:mediator,-:joint mediator
Coefficients: estimated coefficients;
P-Value 2:Tests of relationship with the Predictor


> summary(data.m.m.c.e2.2,only=TRUE)
Identified as mediators:
[1] "m11" "m12"
Selected as covariates:
[1] "m01" "m09"
Tests:
      P-Value 1.y1 P-Value 1.y2 P-Value 2.x1 P-Value 2.x2
m01        0.000        0.053        0.607        0.533
m09        0.043        0.993        0.678        0.292
m11 *      0.000        0.000        0.653        0.000
m12 *      0.003        0.148        0.003        0.000
----
*:mediator,-:joint mediator
P-Value 1:univariate relationship test with the outcome;
P-Value 2:Tests of relationship with the Predictor
```

In summary, if there are multiple outcomes or exposure variables, the "P-Value 1" or "Coefficients" are shown for each of the outcomes and the "P-Value 2" are shown for each of the exposure variables. Significance for the second test is established if the p-value is smaller than "alpha1" for any of the outcomes. Similarly, significance for the first test is established if the p-value is smaller than "alpha2" for any of the exposure variables. Note that the tests are not formal for identifying significant third-variables or covariates. One can set somewhat large significance levels ("alpha1" and "alpha2") to avoid leaving out important variables.

8.3.3   Function *med.big*

The output from the function *data.org.big* is read into the function *med.big* to select third-variables and estimate their effects (indirect effect (IE), total effect (TE), direct effect (DE), and relative effects (defined as IE/TE or DE/TE, denoted as RE)). For example, the following code performs the third-variable analysis on the simulated data set based on the results from *data.org.big*.

```
med.e1<-med.big(data.e1)
```

To perform the analysis, the result of *data.org.big* is not necessary. The arguments for *data.org.big* can be input to *med.big*, then *data.org.big* will be called within *med.big*. The output of the function *med.big* is classified as the "med.big" class and the generic function *print.med.big* shows the estimation results for the "med.big" class.

```
> med.e1$
$y
          pred
de    0.01857787
m161  0.00000000
m162  0.98931416
m11   1.21150833
m12   0.52051526
te    2.73991562
```

That is, the estimated direct effect from x is 0.019. The estimated indirect effect of $m11$ and $m12$ is 1.21 and 0.52, respectively. The results can be interpreted as that on average, when x increases by one unit, y increases by 1.21 and 0.52 units through the change of $m11$ and $m12$, respectively. The estimated indirect effect of $m162$ is 0.99, which is when x changes by 1 unit, y increases by 0.99 unit through the change in the prevalence of $m16$ being $2, p(m16 = 2)$.

The above results are returned with the function *print.med.big*. We can use the results to calculate the estimated relative effect. For example, the following codes print the estimated relative effect:

```
> med.e1.p<-print(med.e1,print=F)
> med.e1.p$results[[1]]/
  med.e1.p$results[[1]][nrow(med.e1.p$results[[1]]),]
          $pred
de    0.006780455
m161  0.000000000
m162  0.361074680
m11   0.442169941
m12   0.189974924
te    1.000000000
```

The *results* element of *med.e1.p* is a list, where the ith item of *med.e1.p\$results* is a matrix that gives the estimated third-variable effects for the ith outcome. The jth column of the matrix is for the jth predictor. The following example shows the results for multiple predictors and multivariate outcomes.

```
> med.m.m.c.e2.2<-med.big(data.m.m.c.e2.2)
> med.m.m.c.e2.2
$y1
           x1         x2
de  0.2494818 0.9580249
m11 0.1204943 0.7289243
m12 0.4290098 0.3444318
te  0.7989859 2.0313811


$y2
           x1         x2
de  0.00000000 0.0000000
m11 0.08861801 0.5360902
m12 0.12734912 0.1022426
te  0.21596713 0.6383329
```

Finally, in the *mmabig* package, the relationships between third-variables and the exposure variable(s) are fitted through generalized smoothing splines to allow the fit of potential nonlinear relationships. In the package, the *ns()* function is used to generate the B-spline basis matrix for a natural cubic spline. The argument "df" in the *med.big* function is used to assign the degrees of freedom for the spline basis matrix. By default, the degree of freedom is 1, which is to fit a linear relationship. In Section 8.5, we give examples to fit nonlinear relationship between third-variables and the exposure variable. We also allow generalized linear models to fit the relationship between outcomes and all predictors in the full model. The argument used in the *med.big* function to define the generalized linear model is "family1". It is a list with the ith item defines the conditional distribution of the ith outcome, $y[, i]$ given the predictors, and the linkage function that links the mean of the outcome with the system component if generalized linear model is used as the final full model. The default value of "family1" is *gaussian(link="identity")* for continuous outcomes, and *binomial(link = "logit")* for binary ones.

8.3.4   Function *mma.big*

Finally, the function *mma.big* uses the bootstrap method to estimate variances and generate confidence intervals for the estimated third-variable effects. The function can be used based on the output of *data.org.big*, or it can start the mediation analysis from scratch. The *summary* function is created to summarize inference results. In the summary function, three different sets of confidence intervals are calculated: based on the normal assumption of bootstrap method (lwbd, upbd), on the quantiles (lwbd_q,upbd_q), and on the confidence ball (lwbd_b, upbd_b) (see Chapter 7). The following codes first simulate a data set and then summarize the result. The *summary* function generates Figure 8.2 showing the estimated relative effects with confidence intervals. The results from the *summary* function are shown after the codes.

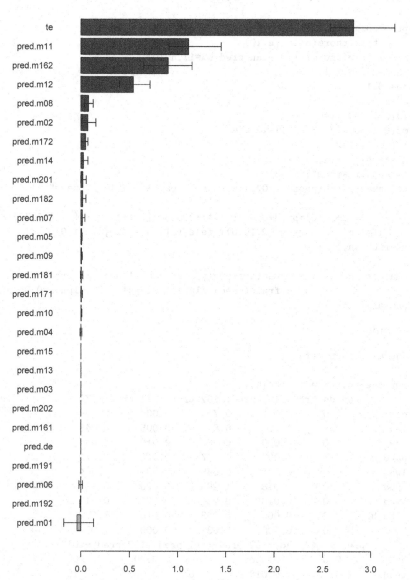

FIGURE 8.2
The figure result from the summary function.

R Codes:

```
set.seed(1)
n=100
pred=rbinom(n,1,0.5)
m1=matrix(rnorm(n*10),n,10)
m2<-matrix(rnorm(n*10,mean=pred,sd=1),n,10)
m3.1=m2[,6:10]
m3=m3.1
m2=m2[,1:5]
m3[m3.1<=0.1]=0
m3[0.1<m3.1 & m3.1<=1]=1
m3[m3.1>1]=2
m3<-apply(m3,2,as.factor)
m<-data.frame(m1,m2,m3)
colnames(m)<-c(paste("m0",1:9,sep=""),paste("m",10:20,sep=""))

lu<--0.5363+0.701*pred+0.801*m[,1]+0.518*m[,2]+1.402*m[,11]+0.773*m[,12]+
    ifelse(m[,16]=="2",2.15,0)+ifelse(m[,16]=="1",0.201,0)
y<-rnorm(n,lu,1)

mma.e1<-mma.big(x=m,y=data.frame(y),mediator=1:ncol(m),pred=
               data.frame(pred),alpha=1,alpha1=0.05,alpha2=0.05,n2=3)
summary(mma.e1)
```

Results:

The mediaiton effects:

For the response variable,y,

	pred.de	pred.m161	pred.m162	pred.m171	pred.m172
est	0	0.018	0.748	0.000	0.069
mean	0	0.000	0.898	0.005	0.048
sd	0	0.000	0.264	0.010	0.041
upbd_q	0	0.000	1.147	0.016	0.072
lwbd_q	0	0.000	0.645	-0.002	0.004
upbd	0	0.018	1.265	0.020	0.151
lwbd	0	0.018	0.231	-0.020	-0.012
upbd_b	0	0.000	0.903	0.016	0.072
lwbd_b	0	0.000	0.631	0.000	0.072

	pred.m181	pred.m182	pred.m191	pred.m192	pred.m201
est	0.008	0.000	0	0.000	0.000
mean	0.005	0.018	0	-0.003	0.019
sd	0.013	0.032	0	0.006	0.033
upbd_q	0.019	0.052	0	0.000	0.055
lwbd_q	-0.005	0.000	0	-0.010	-0.002
upbd	0.035	0.062	0	0.012	0.065
lwbd	-0.018	-0.062	0	-0.012	-0.065
upbd_b	0.020	0.055	0	0.000	0.001
lwbd_b	0.000	0.000	0	0.000	-0.002

	pred.m202	pred.m01	pred.m02	pred.m03	pred.m04
est	0	0.112	-0.016	0	0.000
mean	0	-0.039	0.072	0	0.000
sd	0	0.166	0.079	0	0.010
upbd_q	0	0.133	0.154	0	0.011
lwbd_q	0	-0.175	0.010	0	-0.009
upbd	0	0.438	0.139	0	0.021
lwbd	0	-0.213	-0.170	0	-0.020
upbd_b	0	0.144	0.160	0	0.011
lwbd_b	0	-0.082	0.008	0	-0.001

	pred.m05	pred.m06	pred.m07	pred.m08	pred.m09
est	0.000	-0.003	0.008	0.037	0.000
mean	0.008	0.000	0.015	0.079	0.007
sd	0.007	0.020	0.024	0.047	0.007
upbd_q	0.013	0.018	0.041	0.128	0.014
lwbd_q	0.000	-0.020	0.000	0.039	0.002
upbd	0.014	0.036	0.055	0.129	0.013
lwbd	-0.014	-0.042	-0.040	-0.056	-0.013
upbd_b	0.014	0.019	0.043	0.069	0.006
lwbd_b	0.000	-0.021	0.000	0.038	0.002

	pred.m10	pred.m11	pred.m12	pred.m13	pred.m14
est	0.000	1.211	0.476	0	0.000
mean	0.003	1.111	0.539	0	0.025
sd	0.005	0.315	0.168	0	0.044
upbd_q	0.009	1.448	0.712	0	0.072
lwbd_q	0.000	0.908	0.399	0	0.000
upbd	0.011	1.829	0.804	0	0.086
lwbd	-0.011	0.593	0.147	0	-0.086
upbd_b	0.001	1.474	0.723	0	0.076
lwbd_b	0.000	0.906	0.501	0	0.000

	pred.m15	pred.te
est	0	2.670
mean	0	2.810
sd	0	0.405
upbd_q	0	3.243
lwbd_q	0	2.571
upbd	0	3.464
lwbd	0	1.875
upbd_b	0	3.278
lwbd_b	0	2.581

Or we can use the output from *data.org.big* and in the *summary* function, we can use the argument "RE=T" to show the inference result for relative effects. The following is an example.

```
> mma.e1.2<-mma.big(data=data.e1.2,alpha1=0.05,alpha2=0.05,n2=3)
> summary(mma.e1.2,RE=TRUE,plot=F)
The relative effects:

For the response variable,y,
```

	pred.de	pred.m161	pred.m162	pred.m11	pred.m12
est	0.007	0.000	0.361	0.442	0.190
mean	0.058	0.004	0.307	0.458	0.173
sd	0.100	0.005	0.051	0.112	0.035
upbd_q	0.165	0.009	0.355	0.551	0.198
lwbd_q	0.000	0.000	0.259	0.342	0.137
upbd	0.203	0.009	0.461	0.662	0.258
lwbd	-0.189	-0.009	0.262	0.222	0.122

8.3.5 Generic Functions

The *mmabig* package provides generic functions to help explain results from the third-variable effect analysis. As has shown, the results from *data.org.big* can be summarized to identify potential third-variables and covariates, and show test results (p values) for each association of interests. Also, the outputs from the function *mma.big* can be summarized to show the inference results on third-variable effects (estimates, standard deviations and confidence intervals).

Further, the function *joint.effect* makes inferences on the joint effect of multiple third-variables from the *mmabig* object created by *mma.big*. The following codes show how to get the estimated effect (and relative effect) of the multi-categorical variable "m16", and the joint effect of variables "m11" and "m12".

```
> joint.effect(mma.e1.2,vari=c("m16"))
$variables
[1] "m161" "m162"

$effect
$effect$y
          pred
est    0.9893142
mean   0.8928579
sd     0.3087729
upbd_q 1.1082313
lwbd_q 0.5641390
upbd   1.5944980
lwbd   0.3841304


$relative.effect
$relative.effect$y
           pred
est    0.36107468
mean   0.31088949
sd     0.04993564
upbd_q 0.35565175
```

```
lwbd_q 0.26124579
upbd   0.45894673
lwbd   0.26320263


> joint.effect(mma.e1.2,vari=c("m11","m12"))
$variables
[1] "m11" "m12"

$effect
$effect$y
            pred
est    1.7320236
mean   1.7410628
sd     0.4127931
upbd_q 2.1815458
lwbd_q 1.4631307
upbd   2.5410833
lwbd   0.9229639


$relative.effect
$relative.effect$y
            pred
est    0.6321449
mean   0.6313314
sd     0.1436099
upbd_q 0.7387542
lwbd_q 0.4796777
upbd   0.9136151
lwbd   0.3506746
```

The graphic function, *plot*, helps researchers visualize the complicated relationships among variables and explain the directions of third-variable effects. The following graph shows the third effect of "m11" on the outcome by using the code *plot(mma.e1.2,vari="m11")*.

The upper panel of Figure 8.3 is the boxplot of the coefficients of "m11" in predicting "y" in elastic net regularized regression with the bootstrap samples. All positive coefficients mean that there is a positive relationship between "m11" and "y". The lower panel is the boxplot of the ΔM from the bootstrap samples. All positive ΔM means that when the predictor changes, the average changing rate in "m11" is also positive. That is when the predictor increases value, "m11" increases, in turn, "y" increases. Therefore, "m11" can explain part of the predictor-outcome relationship.

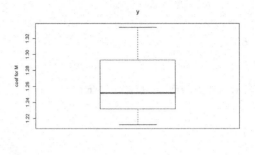

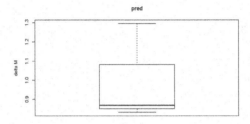

FIGURE 8.3
The figure plotted by the function *plot(mma.e1.2,vari="m11")*.

Figure 8.4 is the graph for the categorical variable "m16". Note that "m16" has three levels. Therefore two binary variables "m161" and "m162" are created for the regressions. Figure 8.4 shows the relationships of each of the generated binary variables.

8.3.6 Call *mmabig* from SAS

As for the R package *mma*, we create SAS macros for the *mmabig* package, so that the mediation analysis with high-dimensional data can be analyzed in the SAS environment. The use of SAS macros is similar to those for the *mma* package. The parameters in the SAS macro templates are to be modified according to needs. All the SAS macros are provided through the online supplementary material.

8.4 Sensitivity and Specificity Analysis

A series of simulations are designed to check the sensitivity and specificity of the proposed method in identifying important third-variables. The purpose

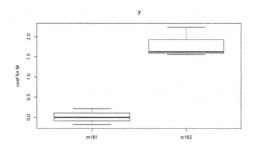

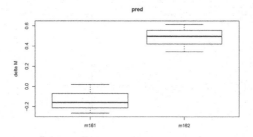

FIGURE 8.4

The figure plotted by the function *plot(mma.e1.2,vari= "m16")*.

is to find out how the power and the misclassification rate are influenced by factors such as the number of variables, the sample size, the types of the third-variable (continuous or categorical) and the effect size. The original data are generated as follows:

$$
\begin{aligned}
x_i &\sim N(0,1), i = 1, \ldots, n; \\
m_{ij} &\sim N(0,1), j = 1, \ldots, J - 2; \\
m_{ij} &\sim N(\alpha_1 x_i, 1), j = J - 1, J; \\
y_i &\sim N(\beta_0 x_i + \beta_1 m_{i1} + \beta_1 m_{iJ}, 1), i = 1, \ldots, n.
\end{aligned}
\qquad (8.5)
$$

Figure 8.5 shows how the power changes with different factors. The power is defined as the proportion of times that the 95% confidence interval of the average indirect effect does not include 0. For each scenario, we simulate the dataset 100 times and calculate the proportion of times that the important mediator m_J is successfully identified. In Panel A, the small dash lines show how the power changes with α_1 when β_1 is set at 1. In comparison, when α_1 is set at 1, the solid lines show how the power changes with β_1. Note that when one of the values of β_1 or α_1 is fixed at 1, the indirect effect of M_J is α_1 or β_1, respectively, since theoretically, the actual indirect effect is $\alpha_1 \times \beta_1$.

The different sample sizes, n, are shown by different colors: black, red, green, purple, and blue represent n at $100, 150, 200, 250$ and 300, respectively.

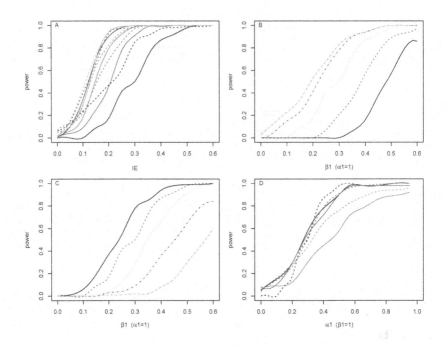

FIGURE 8.5
Sensitivity in identifying important third-variables. For Panel A, solid lines are when $\alpha = 1$ and the small dash lines are for $\beta = 1$. Black, red, green, dark blue, and light blue represent n at 100, 150, 200, 250 and 300, respectively. Solid, small dashes, dot, dot-dash, and dash lines represent $\beta_0 = 0.1, 0.3, 0.5, 0.7, 0.9$, respectively for Panel B, and $J = 10, 20, 30, 40, 50$ for Panel C. For Panel D, the solid black line represents the power for the binary third-variable at $c_1 = 0$, while small dashed black line represents power for the continuous third-variable. The red lines are for $|c_1| = 0.5$ and green for $|c_1| = 1$, where solid is for positive and small dash is for negative c_1s.

In general, we found that at the same sample size and effect size, fixing β_1 at 1 while changing α_1 is related with a little higher power than changing α_1 but fixing β_1 at 1. That is, at the same effect size, the coefficient of the third-variable in predicting the outcome (the β) has a bigger influence on the power than the coefficient of the exposure variable in predicting the third-variable (the α). This is due to the fact that in estimating α_1, the regularization method is not used, therefore, the effect of α is not shrunk. We also found that when sample size increases, the difference in power by coefficients (α or β) becomes undetectable.

Panel B of Figure 8.5 shows how the power changes with β_0, the intercept in predicting y, when α_1 is fixed at 1 and β_1 changes. Black solid, red dashed, green dotted, purple dot-dash, and blue long-dash represent β_0 at 100, 150, 200, 250 and 300, respectively. We see that as β_0 increases, the power of detecting the indirect effect also increases.

Panel C shows how the power changes with the number of potential third-variables, J. Black solid, red dashed, green dotted, purple dot-dash, and blue long-dash represents J at 10, 20, 30, 40 and 50, respectively. As J increases, the power decreases as expected.

Finally, Panel D is to check how the type of third-variables can influence the power of detecting an indirect effect. To create binary third-variables, for $j = J - 1$ or J, we let the new third-variable $\tilde{m}_{ij}$ be 0 if m_{ij} is less than $c1$, otherwise be 1. Changing $c1$ alters the original distribution of m_j by changing the probability $p(m_{ij} = 1)$. The solid black line represents the power for the binary third-variable when $c1 = 0$, while small dashed black line represents power for the continuous third-variable. The red lines are for $|c1| = 0.5$ and green for $|c1| = 1$, where solid line is for positive and dash line is for negative $c1$s. We found that first, when α_1 is small, the power is higher to identify the binary third-variable than the continuous one (solid vs. dash black lines, respectively). However, when α_1 becomes larger, it is easier to identify the continuous third-variable. This is because the indirect effect for the binary third-variable is roughly $E[P(m = 1|x + 1) - P(m = 1|x)]\beta_1$ versus $\alpha_1\beta_1$ for the continuous third-variable. Second, we found that as c_1 moves away from 0, $P(m = 1)$ moves away from 0.5, i.e. the original distribution of m becomes more uneven, hence the power of detecting the important binary third-variable reduces.

Figure 8.6 shows the misclassification rates, $(1 - specificity)$, at the same scenario as in Figure 8.5. The left panel, false positive rate 1, shows the probabilities of taking m_1 as an important third-variable (where $\alpha_1 = 0$ while $\beta_1 \neq 0$) and the right panel, the false positive rate 2, is for identifying m_{J-1} (where $\beta_1 = 0$ while $\alpha_1 \neq 0$) as an important third-variable. Panel A of Figure 8.6 shows that the false positive rate 1 is generally bigger than the false positive rate 2. In addition, the false positive rates increase as the sample size increases. In Panel B, we found that both false positive rates increase as β_0 increases. Panel C shows that the specificity decreases with the number of potential third-variables. And panel D shows that the false positive rates do not change a lot with the original distribution of $P(m = 1)$. The false positive rate for m_1 is zero for different α_1s when m_J is binary. Overall, the type I error is controlled under 0.05.

8.5 Simulations to Illustrate the Use of the Method

The purpose of simulation studies in this section is to illustrate the proposed third-variable (TV) analysis for high-dimensional dataset when there

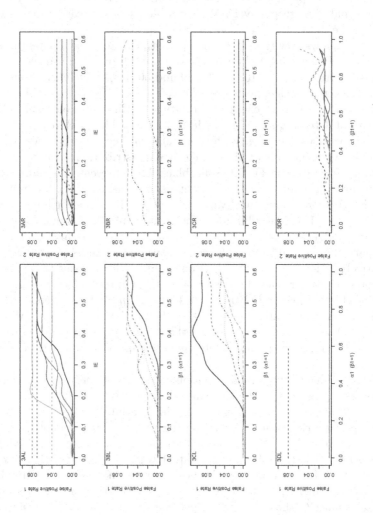

FIGURE 8.6

1-specificity in identifying third-variables. The left column, the false positive
rate 1, shows the probabilities of taking m_1 as an important third-variable,
and the right column, the false positive rate 2, for picking m_{J-1}. For Panel As,
black, red, green, dark blue, and light blue represent n at 100, 150, 200, 250 and
300, respectively. Solid, small dashes, dot, dot-dash, and dash lines represent
$\beta_0 = 0.1, 0.3, 0.5, 0.7, 0.9$, respectively for Panel Bs, and $J = 10, 20, 30, 40, 50$
for Panel Cs. For Panel Ds, the solid black line represents the power for the binary third-variable at $c_1 = 0$, while small dashed black line represents misclass-
ification rate for the continuous third-variable. The red lines are for $|c_1| = 0.5$
and green for $|c_1| = 1$, where solid is for positive and small dash is for negative
c_1s.

are nonlinear relationships or colinearities among variables. In the following, we demonstrate the method in different situations.

8.5.1 $X - TV$ Relationship is Nonlinear

The data are simulated similarly as equations (8.5) except that the exposure variable is simulated with a mean of c_2, i.e., $x_i \sim N(c_2, 1)$. In addition, when $j = J - 1$ or J, m_{ij} is simulated by

$$m_{ij} \sim N(\alpha_1 x_i^2, 1). \qquad (8.6)$$

That is, the exposure-third-variable relationship is quadratic. Theoretically, if a linear model is forced to fit the relationship between x and m_J, the coefficient for x would be insignificantly different from 0. Therefore m_J is not going to be identified as an important third-variable. A smoothing spline to model the relationship between x and third-variables can solve the problem. The following R codes give the example of generating the simulation data, and fit a cubic spline for using the exposure variable to predict the third-variable.

```
set.seed(t1)
pred=rnorm(n)
m1=matrix(rnorm(n*(J-2)),n,J-2)
m2<-matrix(rnorm(n*2,mean=alp1*pred,sd=1),n,2)
m2<-ifelse(m2>c1,1,0)
m2<-apply(m2,2,as.factor)
m<-data.frame(m1,m2)
colnames(m)<-c(paste("m0",1:9,sep=""),paste("m",10:J,sep=""))
lu<-beta0*pred+beta1*ifelse(m[,1]==1,1,0)+
    beta1*ifelse(m[,J]==1,1,0)
y<-rnorm(n,lu,1)
mma.sim1<-mma.big(x=data.frame(m),y=data.frame(y),
                  mediator=1:ncol(m),pred=data.frame(pred),
                  alpha1=0.05,alpha2=0.05,n2=100,df=3)
```

Figure 8.7 shows the simulation results with $c_2 = 0$, $J = 20$ and $n = 100$. The solid line indicates the power of identifying M_J when $E(M_J|x) = \alpha_1 x$, i.e. when M_J has a linear relationship with x. When the third-variable has a quadratic relationship with x as in Equation (8.6), the small dash line is the power of identifying it using a linear model to fit the $x - tv$ relationship, and the dot-dash line is the power for using a smoothing spline.

If a linear model is fitted for the $x - tv$ relationship, as c_2 moves to the right of 0, a positive coefficient for x becomes more likely to be positive. If c_2 moves to the left, a negative coefficient is more likely to be fitted for x. Therefore the important third-variable can be found with more power but with a misleading explanation of the direction of the indirect effect. Using the smoothing spline to fit the relationship, the actual $E(\Delta M_j)$ is $2\alpha \times c_2 + \alpha$. c_2 still influences the power of detecting m_J, but in a very different way.

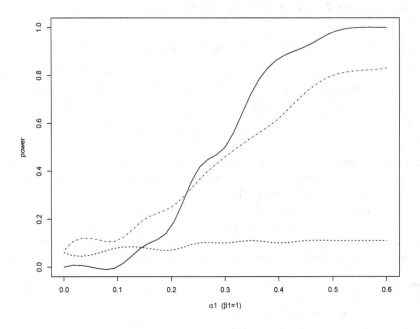

FIGURE 8.7

The power of identifying a third-variable that is linearly related with x (solid), quadratic related and fitted with linear model (small dashes), quadratic related and fitted with smoothing spline (dot-dashes).

Figure 8.8 shows how the power changes with c_2 using the linear regression (left) or smoothing spline (right). The dark blue, light blue, green, red, and black represents $|c2|$ be $0, 0.25, 0.5, 0.75$ and 1, respectively. Solid lines are for positive c_2 and small dashes are for negative c_2. We see that for the linear regressions, the power increases as c_2 moves away from 0 almost symmetrically. For smoothing splines, the power is the smallest when $c_2 = -0.5$, at which $E(\Delta M_j)$ is 0. As c_2 moves away from -0.5, the power increases.

Figure 8.9 shows the interpretation of the third-variable effect when $\alpha_1 = 1$ and $c_2 = 0$. The upper panel shows the boxplots of the coefficients for m_{20}^* from the 100 repeats. The lower panel shows the estimated $\Delta M_{i,20}$ vs. x_i from the first iteration. For the left panel, ΔM is calculated using linear models, and the right using smoothing splines. Smoothing spline method helps to interpret the direction of third-variable effect correctly.

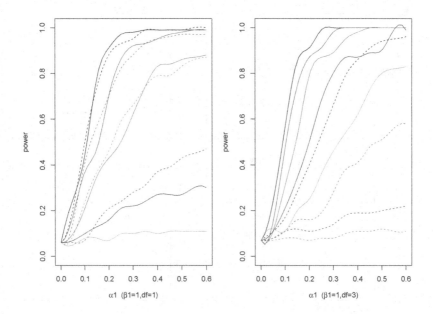

FIGURE 8.8

The power of identifying a third-variable using linear model (left) or smoothing spline (right) as c_2 changes. $c_2 = 0$ has the lowest power. Then the power increases as $|c_2|$ goes $0.25, 0.5, 0.75$ and 1, respectively. Solid lines are for positive and small dashes are for negative c_2.

8.5.2 $TV - Y$ Relationship is Nonlinear

When the $TV - Y$ relationship is nonlinear, transformations of the third-variables are needed to accurately estimate the third-variable effect. Now simulations are generated as in Equations (8.5), except that the outcome is generated in the following way, which includes an additional important third-variable m_J^2:

$$y_i \sim N(\beta_0 x_i + \beta_1 m_{i1} + \beta_1 m_{iJ} + \beta_1 m_{iJ}^2, 1), i = 1, \ldots, n.$$

In such case, m_J^2 is created as a $(J+1)^{th}$ third-variable. The following codes give an example of how to generate the dataset and the third-variable analysis.

```
set.seed(t1)
pred=rnorm(n)
m1=matrix(rnorm(n*(J-2)),n,J-2)
m2<-matrix(rnorm(n*2,mean=alp1*pred,sd=1),n,2)
```

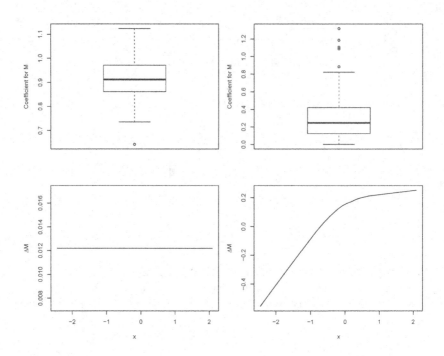

FIGURE 8.9

The upper panel shows the boxplots of the coefficients for m_{20}^* from the 100 repeats. The lower panel shows the average estimated ΔM with x. For the left panel, ΔM is calculated using linear model, and the right with smoothing splines.

```
m<-data.frame(m1,m2,m2[,2]^2)
colnames(m)<-c(paste("m0",1:9,sep=""),paste("m",10:(J+1),sep=""))
lu<-beta0*pred+beta1*m[,1]+beta1*m[,J]+beta1*m[,J+1]
y<-rnorm(n,lu,1)
mma.sim1<-mma.big(x=m,y=data.frame(y),mediator=1:ncol(m),
            pred=data.frame(pred),alpha1=0.05,
                  alpha2=0.05,n2=100,df=3)
summary.sim1=summary(mma.sim1)
```

Figure 8.10 shows the power of identifying m_J (solid) and m_J^2 (small dashes) as an important third-variable separately. Note that the power is defined as the proportion of times that the 95% confidence interval of the average indirect effect does not include 0. The power of identifying m_J is higher than of identifying m_J^2. Since the effect of the Jth variable is most important,

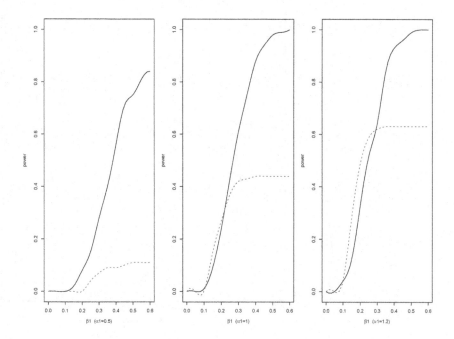

FIGURE 8.10
The power of identifying a linear third-variable (solid) and a quadratic third-variable (dashes).

we can use the *joint.effect* function to find the effect of m_J and m_J^2. The code example is `joint.effect(mma.sim1,vari=c("m10","m11"))`.

We have $E(m_J^2|x) = \alpha_1 x^2 + 1$, which decreases with x when it is negative. Therefore, when x is negative, the indirect effect from m_J^2 is negative, while the indirect effect is positive when x is positive. When x is centered at 0, the average indirect effect is around 0 even when m_J^2 is an important third-variable. Therefore, the power looks to be smaller for m_J^2 than for m_J. When we shift x away from 0, as is shown in Figure 8.11, the power of identifying m_J^2 becomes larger. In Figure 8.11, the four graphs show the powers when x is centered at $-1, -0.5, 0.5$ and 1 separately from left to right.

8.5.3 When Third-Variables are Highly Correlated

The last simulation study is to compare the proposed method with the traditional coefficient product (CP) method. We compare the power of detecting important third-variables when third-variables are correlated. In this case, we

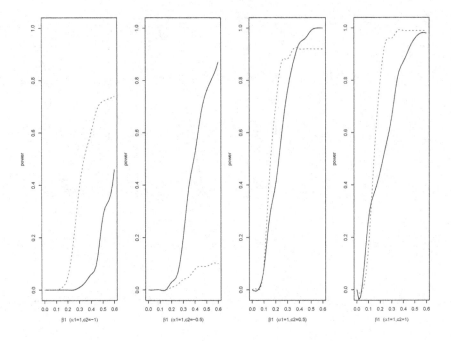

FIGURE 8.11

The power of identifying a linear third-variable (solid) and a quadratic third-variable (dashes) when the center of x shifts among $\{-1, -0.5, 0.5, 1\}$ from left to right.

still use equations in (8.5) to simulate the original data. In addition, we set $J = 10$ and $n = 200$. However, in generating the third-variables, we put an autocorrelation structure among the third-variables such that the correlation coefficient between m_{j_1} and m_{j_2} is $\rho^{|j_1 - j_2|}$. The following R codes generate the data for simulation.

```
set.seed(t1)
pred=rnorm(n)
mean1=rep(0,J/2)
var1=auto.var(J/2,rho)
m1=mult.norm(mean1,var1,n)
m2=mult.norm(mean1,var1,n)
m2=m2+alp1*pred
m<-data.frame(m1,m2)
colnames(m)<-c(paste("m0",1:9,sep=""),paste("m",10:J,sep=""))
lu<-beta0*pred+beta1*m[,1]+beta1*m[,J]
```

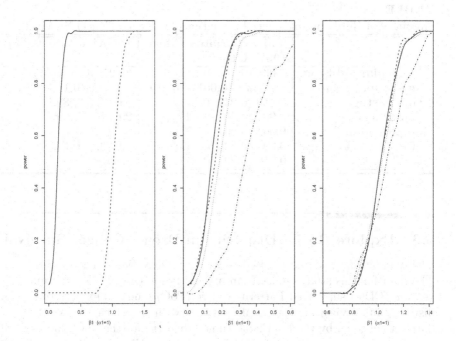

FIGURE 8.12
The left panel shows the powers in detecting m_{10} using the mmabig method (solid line) and the CP method (small dashes). The middle (mmabig) and right (CP) panels show how the power changes with ρ, where ρ changes among 0 (solid), 0.3 (small dashes), 0.6 (dots) and 0.9 (dot-dashes).

```
y<-rnorm(n,lu,1)
mma.sim1<-mma.big(x=m,y=data.frame(y),mediator=1:ncol(m),
            pred=data.frame(pred),alpha1=0.05,
                alpha2=0.05,n2=100)
summary.sim1=summary(mma.sim1)
```

Figure 8.12 shows the comparison results. The left panel shows the comparison of powers in detecting m_{10} using the proposed method (solid line) and the CP method (small dashes). We found that the proposed method has a much larger power. The middle and right panels show how ρ influences the power by the proposed method (mmabig) and the CP method, respectively. In the figure, ρ changes among 0 (solid), 0.3 (small dashes), 0.6 (dots), and 0.9 (dot-dashes). Note that to show the figures better, we change the range of the indirect effect for our method to $(0, 0.6)$ and for CP to $(0.8, 1.4)$. We found the pattern that power decreases with ρ for both methods.

TABLE 8.1
Summary of third-variable effect Estimations for Breast Cancer Survival.

	Third-Variable Effect (95% CI)	Relative Effect (%)
molecular subtype	0.064 (0.005, 0.144)	30.27 (3.15, 56.33)
age at diagnosis	−0.033 (−0.001, −0.070)	−15.9 (−0.1, −43.0)
tumor stage	0.115 (0.041, 0.189)	54.9 (25.3, 89.5)
tumor size	0.062 (0.003,0.121)	29.5 (5.4, 53.6)
employment	0.004 (0.001, 0.051)	1.3 (0.1, 8.4)
Hormonal Therapy	0.008 (0.001, 0.052)	3.0 (0.1, 8.3)
total effect	0.210 (0.097, 0.394)	100.0

8.6   Explore Racial Disparity in Breast Cancer Survival

The mmabig method can be used to explore disparities in health outcomes. The data for this study comes from a Centers for Disease Control and Prevention (CDC) supported Pattern-of-Care (PoC) study to explore the racial disparity in survival among female patients with breast cancer. We use the data set collected by the Louisiana Tumor Registry for the PoC study, which re-abstract medical records of 1453 Louisiana non-Hispanic white and black women diagnosed with invasive breast cancer in 2004. We found that the 3-year death rate of breast cancer is much higher for blacks than for whites (odds ratio=2.03, [79]). To explore the racial disparity in mortality, multivariate mediation analysis can be used. In this case, the exposure variable x is the binary race (black or white). The outcome of this analysis is time-to-event, where y_i denotes the observed time (either censoring or death time) for subject i and c_i be the indicator that the time corresponds to an event (i.e. $c_i = 1$ for death and $c_i = 0$ for censoring). We consider 25 variables as potential third-variables that include demographic information of the patients, residence environment factors, tumor characteristics at the diagnosis, and treatment information.

A cox model is used to build the hazard function of death in breast cancer. We first transform the third-variables to have a rough linear relationship with the outcome and then apply the proposed mediation analysis to the transformed data through the *mmabig* package. Table 8.1 shows the selected third-variables and the estimates of third-variable effects in explaining the racial disparity in breast cancer survival. In the table, relative effect is defined as the third-variable effect divided by the total effect. Relative effect is explained as that if the corresponding third-variable can be manipulated to distribute equally among blacks and whites, the potential change in the racial disparity in survival.

Table 8.1 shows that the significant third-variables include the molecular cancer subtype, age at diagnosis, tumor stage, tumor size, employment status,

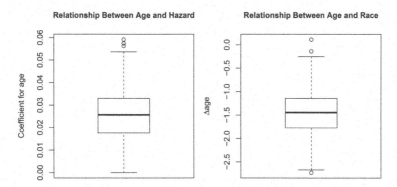

FIGURE 8.13

The left panel is the boxplot of the coefficients for age in the cox model through the bootstrap samples. The right panel shows the average ΔM for age.

and hormonal therapy status. The exposure variable, race, is not significant, meaning that third-variables considered in the study can completely explain the racial disparity in breast cancer survival.

The relative effect of age at diagnosis is negative, which can be explained by Figure 8.13. The left panel is the boxplot of the coefficients for age in the cox model through the bootstrap samples. The coefficients are mostly positive, meaning that the hazard rate increases with age at diagnosis. The right panel shows the average ΔM for age, which is negative, meaning that compared with whites, blacks were diagnosed at an average younger age. Therefore, if the age at diagnosis can be manipulated so that the age distributions are equivalent among blacks and whites, the racial disparity in breast cancer survival would increase by 15.9%. Rather than explain the disparity, adjust for the age would enlarge the disparity. Thus the relative effect is negative. The R package *mmabig* provides the plot function to generate similar pictures for each potential third-variables in explaining the directions of third-variable effects.

If all variables are included for exploration, we randomly selected bootstrap samples 1000 times and repeat the proposed method on the bootstrap samples. We show all important third-variables that were selected at least half of the times and order them by the numbers of times of being selected (i.e. the number of times that the coefficient of the third-variable is different from 0): metastasis at diagnosis (998), number of lymph nodes (995), size of tumor (884), HER2 status(747), marital status(686), hospital Commission on Cancer status (625), privately owned hospital (601), age at diagnosis (510). We found that after adjusting for these third-variables, all racial difference was explained. That is, the coefficients for race are mostly shrunk to 0 for the 1000 bootstrap samples.

9

Interaction/Moderation Analysis with Third-Variable Effects

Moderation effect refers to the interaction effect of two or more variables on the outcome. Within the settings of third-variable effect analysis (TVEA), the interaction effect can be between the moderator and the exposure, where the relationship between the exposure and outcome (direct moderation effect) and/or the relationship between the exposure and a third variable are different at different levels of the moderator. In addition, the interaction effect can be between the moderator and any third-variables such that the third-variable and outcome relationship is different at various level of the moderator(see Figure 9.1). One of the main purposes of moderation analysis within TVEA is to make inferences on third-variable effects at different levels of the moderator. This chapter is derived in part from an article published in the Journal of Applied Statistics [88] in August 2021, available at https://doi.org/10.1080/02664763.2021.1968358.

Moderation has been widely used in the psychological research, behavior science, and public health studies. Recently many research methods have been developed to make inferences on moderation effects within the third-variable effect analysis [15, 18, 34]. However, there are still many challenges in the field such as automatically identification of interaction terms and the dealing of potential nonlinear relationships. In this Chapter, we implement the moderation analysis within the generalized third-variable analysis method that we built up in the book.

The Multiple Additive Regression Trees (MART) [83, 26] are used in the TVEA to perform the moderation analysis. Because of the tree structure of MART, interaction effects among variables are allowed and selected automatically through the model building process. Therefore, the moderation effects can be estimated without manually creating interaction terms which are then put into the predictive model. In addition, nonlinear relationship among variables is allowed with the predictive models MART and smoothing splines. To implement the method, the R package *mma* is further developed to make inferences on moderation effects. The package can be used to make inferences on third-variable effects with or without moderation. It also provides visual aids to help understand the trend and direction of associations among variables.

We develop a method to make inferences on moderation effect in Section 9.1. The algorithm and R package are also discussed in the section. We

DOI: 10.1201/9780429346941-9

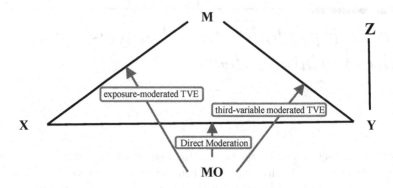

FIGURE 9.1
Moderation diagram. X is the exposure variable, M is a third variable, Y is the outcome, MO is a moderator and Z is the vector of other covariates.

then illustrate the proposed method through simulations at different moderation scenarios in Section 9.2. In Section 9.3, the method is applied to explore the trend of racial disparities in the Oncotype DX usage among breast cancer patients. Oncotype DX® (ODX) is a genomic test that can differentiate breast cancer patients by the risk of recurrence to project prognosis and chemotherapy benefit. ODX test is based on 21-gene expression levels and it produces a recurrence score, a number between 0 and 100. A raised ODX score indicates a higher probability of cancer recurrence and more benefit from chemotherapy. We propose to explore the trend of racial disparities in ODX exam from 2010 to 2015 using a moderation effect inference with TVEA. In this case, the exposure variable is the race (non-Hispanic blacks vs. non-Hispanic whites), the outcome is the use of ODX test (yes/no) and third-variables are all variables that can explain the racial difference in ODX usage. The moderator is the diagnosis year.

9.1 Inference on Moderation Effect with Third-Variable Effect Analysis

9.1.1 Types of Interaction/Moderation Effects

As discussed in previous Chapters, the third-variable effects mainly include direct and indirect effect. The direct effect refers to the causal or non-causal relationship between the exposure and the outcome after adjusting for other variables. The indirect effect of a third variable is the effect from exposure to the third-variable to the outcome.

It is not uncommon that the relationship between an exposure variable and an outcome changes at different levels of an interactive variable. We usually call the interactive variable a moderator. A moderation effect is basically the interaction effect of the moderator with other predictors on the response variable(s). In TVEA, the moderator takes into effect in three basic forms: exposure-moderated third-variable effect (TVE), third-variable moderated TVE, and direct moderation effect as is shown in Figure 9.1.

- *The exposure-moderated TVE* is that at different level of the moderator (MO), the effect of the exposure (X) on the third-variable (M), is different.

- *The third-variable moderated TVE* indicates that the interaction effect is between the third-variable and the moderator on the outcome (Y).

- *The direct moderation* is that the exposure and the moderator are interactively related to the outcome.

Usually to make inferences on moderation effects with regression models, we create interaction terms that potentially have the moderation effect and put the terms in predictive models. The test of significant moderation effect depends on the format of the moderation effect. For the exposure-moderated TVE, a significant moderation effect is established when there is a significant interaction effect of the exposure and the potential moderator on the third-variable when covariates are adjusted. A third-variable moderated TVE is established when there is a significant interaction effect of the third-variable and the potential moderator on the outcome when all other variables are adjusted. To establish a direct moderation, there should be an interaction effect of the exposure and the potential moderator on the outcome [33]. Since moderation effect involves interactions, it is important that TVEs be explained at different levels of the moderator. That is, hierarchical models should be interpreted in terms of the highest-order effect.

We propose a moderation effect analysis method that can automatically select important interactive terms and make inferences on moderation effects. The method is an extension to the general TVEA method as described in previous Chapters. It adopts properties of multiple additive regression trees (MART). We have also revised the R package *mma* to incorporate the proposed moderation analysis.

9.1.2 Moderation Effect Analysis with MART

MART is a special case of the generic gradient boosting approach on trees [22]. The MART method has been introduced in Chpater 3. In general, MART employs two algorithms: classification and regression trees (CART) [8] and boosting which builds and combines a collection of models. Inheriting the benefits from both algorithms, MART has been proved to have excellent predictive accuracy (e.g. [83] and [26]). Due to the hierarchical structure of the tree method, MART can pick up important interaction effect among variables without the necessity to include specific interaction terms. In addition, MART can model potential nonlinear relationships among variables without transforming those variables. We use these special properties of MART in the general TVEA for inferences on moderation effects. The following Algorithm outlines steps for making inferences on moderation effect.

Algorithm 9 Algorithm for Moderation Analysis
Assume we have the observations $(y_i, x_i, M_{1i}, \ldots, M_{Pi})$, for $i = 1, \ldots, n$. Let $D_x = \{x_i | x_i \in domain_{x*}\}$ and N be a large number.

> *1. Fit a MART on the response variable, y_i, where the predictors are the exposure variable (x_i), all potential third variables $(M_{1i}, \ldots, M_{pi})$, and the moderator (denote as MO_i) such that*
>
> $$E(Y_i) = f(x_i, M_{1i}, \ldots, M_{Pi}, MO_i), \quad for \ i = 1, \ldots, n,$$
>
> (9.1)

> *2. Fit joint models where the exposure variable, moderators, and other covariates are used to predict the third-variables M_p, $p = 1, \ldots, P$. Ignoring the covariates for now, the fitted models have the following format:*
>
> $$\left(\begin{pmatrix} l_1(M_{1i}) \\ l_2(M_{2i}) \\ \ldots \\ l_P(M_{Pi}) \end{pmatrix} \middle| x_i \right) \sim \Pi \left(\begin{pmatrix} g_1(x_i, MO_i) \\ g_2(x_i, MO_i) \\ \ldots \\ g_P(x_i, MO_i) \end{pmatrix}, \Sigma \right),$$
>
> (9.2)

> *where l_ps are linking functions that link each third-variable with the fitted models g_p, so that we can deal with different format of the third-variables. For the fitted models g_p, we proposed to use smoothing splines, so that nonlinear relationships and low-level interactions can be considered in the model fitting. Π is the joint distribution of $\mathbf{M}$ given X, which has a mean vector $\mathbf{g}(x_i, MO_i)$ and variance-covariance matrix Σ.*

> *3. At each level, k, of the moderator, estimate the third variable effect, $k = 1, \ldots, K$:*

> *(a) To estimate the total effect at level k within the subpopulation where $MO_i = k$:*

> *i. Randomly draw N xs from D_x with replacement, denote that as $\{x_j, j = 1, \ldots, N\}$.*

ii. *Randomly draw* $(M_{1j1}, \ldots, M_{Pj1})^T$ *given* $X = x_j$ *from equation (9.2) for* $j = 1, \ldots, N$.

iii. *Randomly draw* $(M_{1j2}, \ldots, M_{pj2})^T$ *given* $X = x_j + a$ *from equation (9.2) for* $j = 1, \ldots, N$.

iv. *The total effect at level* k *of MO is estimated as* $TE_k = \frac{1}{Na} \left[\sum_{j=1}^{N} f(x_j + a, M_{1j2}, \ldots, M_{Pj2}, \right.$

$$MO_i = k) - \sum_{j=1}^{N} f(x_j, M_{1j1}, \ldots, M_{Pj1}, MO_i = k) \bigg].$$

(b) *To estimate the direct effect not through* M_p, *for* $p = 1, \ldots, P$:

i. *Randomly draw* $2N$ M_ps *from the observed* $\{M_{pi}, i = 1, \ldots, n\}$, *with replacement, denote that as* $\tilde{M}_{pj}, j = 1, \ldots, N$.

ii. *Randomly draw* $(M_{1j1}, \ldots, M_{p-1,j1}, M_{p+1,j1}, \ldots, M_{p_j1})^T$ *given* $X = x_j$ *from distribution derived from equation (9.2)* , *where* x_js *were obtained at step 3a3(a)i,* $j = 1, \ldots, N$.

iii. *Randomly draw* $(M_{1j2}, \ldots, M_{p-1,j2}, M_{p+1,j2},$

ldots, $M_{p_j2})^T$ *given* $X = x_j + a$ *from distribution derived from equation (9.2).*

iv. *The direct effect not form* M_p *at level* k *of MO,* $DE_{k, \backslash M_p}$ *is estimated by* $DE_{k, \backslash M_p}$

$= \frac{1}{Na} \left[\sum_{j=1}^{N} f(x_j + a, M_{1j2}, \ldots, M_{p-1,j2}, \right.$

$\tilde{M}_{pj}, M_{p+1,j2}, \ldots, M_{Pj2}, MO_i = k) - \sum_{j=1}^{N} f(x_j, M_{1j1},$

$\ldots, M_{p-1,j1}, \tilde{M}_{p,(N+j)}, M_{p+1,j1}, \ldots, M_{Pj1}, MO_i = k) \big].$

(c) *The average indirect effect of* M_p *at level* k *is* $IE_{k,p} = TE_k - DE_{k, \backslash M_p}$.

The following are some comments for the algorithm:

- If there is an interaction effect of the moderator with the exposure variable or with any of the third-variables on the outcome, the MART modeling should be able to pick up the interaction(s) automatically. By setting a limit to the depth of trees in MART, we can restrict the level of interactions that can be considered. For example, a tree with *depth* = 3 allows for two-way interactions but not three-way or any higher rank interactions.

- The interaction effect of exposure and the moderator on any third-variable is taken care by the resampling of third-variables at each combined level of x and MO.

- The levels of a continuous moderator can be decided according to the analysis purposes. Usually the levels can be chosen by the quartiles/quintiles of the moderator within the original data. Step 3(a)ii in the above algorithm are conducted within the kth interval of the moderator, where the intervals are exclusive and the combination of the K intervals covers the range of the moderator.

- When there are multiple moderators, the moderators are combined to form one moderator where the levels of the combined moderator are any combinations of the levels of the original multiple moderators.

- Bootstrap method can be used to measure the variances of estimated third-variable effects at each level of the moderator.

- The above algorithm is for a single exposure. It is straight forward to extend the algorithm to deal with multi-categorical or multiple exposures: to perform the moderation analysis by building full models on the whole dataset but implementing the general TVEA at different levels of the moderator.

9.1.3   The R Package *mma*

A more recent version of the *mma* (version $8.0 - 0$, published after March 28, 2019) package includes the algorithms for moderation analysis within TVEA. The *mma* package provides generic functions to summarize the inference results of third-variable effect estimates at each level of the moderator. The inference includes the estimates of each third-variable effect (direct/indirect effect), standard deviations and confidence intervals. In addition, plot tools are provided to help explain the direction of third-variable effect that explains the exposure-response relationship, and the change of the TVE at different levels of the moderator. In the following sections, we illustrate in detail on how to use the package for moderation analysis and on how to explain results from the package.

9.2   Illustration of Moderation Effects

In this section, we use simple simulations to show moderation effects at different formats. As shown in Figure 9.1, the moderation effect can be of three formats. We illustrate the effect of the three types of moderation effects through simulations.

9.2.1   Direct Moderation

When the moderator is at different levels, the direct effects of the exposure variable on outcome could be different. This type of moderation is called the direct moderation effect. The effect can be inferred in predictive models by including an interaction term of the exposure variable and the moderator.

In the following simulation, we have one third-variable (m_i), one moderator (mo_i), a predictor $(pred_i)$ and a covariate (c_i). The data are generated in the following way:

$$
\begin{aligned}
c_i, pred_i, mo_i & \overset{ind}{\sim} && N(0,1); i = 1, \dots, n \\
m_i & = && pred_i + \epsilon_{1i}; \\
y_i & = && pred_i + m_i + c_i + mo_i + mo_i \cdot pred_i + \epsilon_{2i};
\end{aligned}
$$

where ϵ_{1i} and ϵ_{2i} are independent random errors with a standard normal distribution. All n are set as 200 in the simulations. The following codes were used to generate the simulation data.

```
library(mma)
a0=0
a1=1
b0=0
b1=1
b2=1
b3=1
b4=1
b5=1
rho=0
mean1=c(0,0)
n=200

#direct moderation: simulation 1
var1=matrix(c(1,rho,rho,1),2,2)
set.seed(1)
d1=mult.norm(mean1,var1,n)

#d1=matrix(runif(2*n),n,2)
c1=rnorm(n)
pred=d1[,1]
mo=d1[,2]
m1=a0+a1*pred+rnorm(n)
y=b0+b1*pred+b2*m1+b3*c1+b4*mo*pred+rnorm(n)
x=cbind(m1,mo,c1)
```

In moderation analysis, we generally report the TVE at different levels of the moderator since a moderation effect is basically an interaction effect. In this simulation, the moderator is numerical. By default, the *mma* package evenly divide the moderator to five quantiles and report the moderation effect at each of the quantiles. Let q_k denote the kth quantile of the moderator. Theoretically, for this simulation, the direct effect at the kth quantile of the moderator should be $1 + E(mo_i|mo_i \in q_k)$. Using the *mma* package, if the linear method is used for the moderation analysis, the interaction term of predictor×moderator needs to be added in the general TVEA as a covariate. If the nonlinear method is used, adding the interaction is not necessary. But if the researcher believes that there is a direct moderation effect, it would be more effective to add the interactive term as a covariate. The following code illustrates how to use the *mma* package to perform the moderation analysis.

```
#nonlinear method
data1<-data.org(x,y,pred=pred,mediator=c(1,3),alpha=0.05,
              alpha2=0.05)
med1=med(data=data1,n=2,nonlinear=T,D=5)
bootmed1<-boot.med(data=data1,n=10,n2=100,nonlinear=TRUE,
              all.model=T,nu=0.001,D=5)
bootmode1=boot.mod(bootmed1,vari="mo",n=10,
              continuous.resolution=5)
summary(bootmode1)
summary(bootmode1,bymed=T)


#linear method
pred=data.frame(pred)
x2=form.interaction(x,pred,inter.cov="mo")
x2=cbind(x,x2)
data2<-data.org(x2,y,pred=pred,mediator=c(1,3),alpha=0.05,
              alpha2=0.05)
med2=med(data=data2,n=2,xmod="mo")
bootmed2<-boot.med(data=data2,n=10,n2=100,all.model=T,
              xmod="mo")
bootmode2=boot.mod(bootmed2,vari="mo",n=10,
              continuous.resolution=5)
summary(bootmode2)
summary(bootmode2,bymed=T)


#nonlinear method with the pred-moderator interaction term
#as a covariate
bootmed1.2<-boot.med(data=data2,n=10,n2=100,nonlinear=TRUE,
              all.model=T,nu=0.001,D=5)
bootmode1.2=boot.mod(bootmed1.2,vari="mo",n=10,margin=0.15,
              continuous.resolution=5)
summary(bootmode1.2,bymed=T)


plot2.mma(bootmode2,vari="m1",moderator="mo")
plot2.mma(bootmode1,vari="m1",moderator="mo")
```

In the above codes for the *summary* function, if the argument *bymed* is set to be true, the third variable effect is shown for each level of the moderator. Figure 9.2 shows the direct effect (DE) of *pred* at different levels of *mo* from MART and from linear regression separately. The *y*-axis is the expected *mo* at each quantile. The *x*-axis gives the estimated direct effect with the confidence interval. Both graphs show an increasing trend of direct effect with the

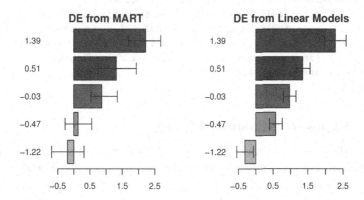

FIGURE 9.2
Direct moderation effect from Simulation 1. The y-axis is denoted by the mean of the moderator at each quantile. The x-axis gives the estimated direct effect and its confidence interval at different levels

moderator as expected. Linear model is more efficient since the simulation was based on linear models.

Each panel of Figure 9.2 can be drawn using the `plot2.mma` function. The following codes detail the codes of generating Figure 9.2.

```
#Figure 9.2: plot the direct effects from linear and
#nonlinear methods
location<-"Figure 2.eps"
postscript(location,width=8,height=4)
par(mfrow=c(1,2),mar=c(1,6,1,1),oma=c(3,2,2,4))
x=summary(bootmode1.2,bymed=T,plot=F)
re<-x$cont.result$results$direct.effect[2,]
#use the quantile interval
upper<-x$cont.result$results$direct.effect[6,]
lower<-x$cont.result$results$direct.effect[7,]
name1<-c(-1.22,-0.47, -0.03, 0.51, 1.39)
bp <- barplot2(re,horiz=TRUE,main=paste("DE from MART"),
        names.arg=name1,plot.ci = TRUE, ci.u = upper,
        ci.l = lower,cex.names=0.9,cex.axis=0.9,las=1,
 beside=FALSE,xlim=range(c(upper,lower)), col=
 rainbow(length(re), start = 3/6, end = 4/6))
x=summary(bootmode2,bymed=T,plot=F)
re=
x$cont.result$cont.resultx$cont.result$results$direct.effect[2,]
#use the quantile interval
upper<-x$cont.result$cont.result$results$direct.effect[6,]
lower<-x$cont.result$results$direct.effect[7,]
```

```
#name1<-colnames(x$cont.result$results$direct.effect)
bp=barplot2(re,horiz=TRUE,main=paste("DE from Linear Models"),
          names.arg=name1,plot.ci = TRUE, ci.u = upper,
          ci.l=lower,cex.names=0.9,beside=FALSE,cex.axis=0.9,
          xlim=range(c(upper,lower)),las=1,col=rainbow(
length(re),start = 3/6, end = 4/6))
dev.off(2)
```

9.2.2   Exposure-Moderated TVE

Next, we illustrate the exposure-moderated TVE, where there is an exposure-moderator interactive effect on the third-variable. The true models are as following:

$$
\begin{aligned}
c_i, pred_i, mo_i &\overset{ind}{\sim} N(0,1); i = 1, \dots, n \\
m_i &= pred_i + mo_i + mo_i \cdot pred_i + \epsilon_{1i}; \\
y_i &= pred_i + m_i + c_i + \epsilon_{2i};
\end{aligned} \tag{9.3}
$$

The codes to simulate the data set:

```
#exposue-moderated TVE: simulation 2
set.seed(2)
d1=mult.norm(mean1,var1,n)

#d1=matrix(runif(2*n),n,2)
c1=rnorm(n)
pred=d1[,1]
mo=d1[,2]
m1=a0+a1*pred+b4*mo*pred+rnorm(n)
y=b0+b1*pred+b2*m1+b3*c1+rnorm(n)
x=cbind(m1,mo,c1)
pred=data.frame(pred)

cova=cbind(mo,form.interaction(x,pred,inter.cov="mo"))
data3<-data.org(x,y,pred=pred,mediator=c(1,3),alpha=0.05,
              alpha2=0.05,cova=cova)
med3=med(data=data3,n=2,nonlinear=T,cova=cova)
```

Using the "mma" package, if the exposure variable is continuous, the relationship between the exposure variable and each potential third-variables adjusting for other covariates is fitted by smoothing splines. Therefore for both linear and nonlinear method, the interaction terms of the exposure and the moderator should be included as covariates to fit the third-variable. The "mma" package provides a function *form.interaction* that can help to form interaction terms between any two variables. For a categorical variable of k levels, the function first transforms the variable into $k - 1$ binary variables and then forms the interactive terms by multiplying each binary variable with the other variable. Thus moderation effect can be inferred for combinations of any type of variables. The following codes show how to fit the model and draw figures to illustrate the direction of moderation effects.

```
#nonlinear method
bootmed3<-boot.med(data=data3,n=10,n2=100,nonlinear=T,
                   all.model=T,nu=0.05,cova=cova)
bootmode3=boot.mod(bootmed3,vari="mo",n=10,cova=cova,
                   continuous.resolution=5,margin=0.4)
summary(bootmode3,bymed=T)

#linear method
med4=med(data=data3,n=2,xmod="mo",cova=cova)
bootmed4<-boot.med(data=data3,n=10,n2=100,all.model=T,
                   xmod="mo",cova=cova)
bootmode4=boot.mod(bootmed4,vari="mo",n=10,cova=cova,
                   continuous.resolution=5)
summary(bootmode4)
summary(bootmode4,bymed=T)

plot2.mma(bootmode4,vari="m1",moderator="mo")
plot2.mma(bootmode3,vari="m1",moderator="mo")
```

The upper panel of Figure 9.3 shows the indirect effect of m from MART and the linear model separately.

The *plot* function delineates how the indirect effect forms from aspects of the exposure-third variable relationship and the third variable-outcome relationship separately at different levels of the moderator variable. Figure 9.4 shows how the relationship between m and *pred* changes at different quantile of *mo*. We see that the slope increases as *mo* increases, which correctly catches the exposure-moderator interaction effect on the third-variable.

Figure 9.4 is generated through the following codes:

```
#Figure 9.4: predictor-third variable relationship$
x=bootmode3
vari="m1"
xlim=NULL
xj=1
location<-"Figure 4.eps"
postscript(location,width=8,height=5)
par(mfrow=c(3,2),mar=c(6,6,1,1),oma=c(3,2,2,4))
for(1 in 1:5){
  if(is.factor(x$a.contx$moder.level$moder))
    temp.all=(x$a.contx$moder.level$moder==
        x$a.contx$moder.level$moder.level[1]
        & !is.na(x$a.contx$moder.level$moder))
  else
    temp.all=x$a.contx$moder.level$levels[,1]
```

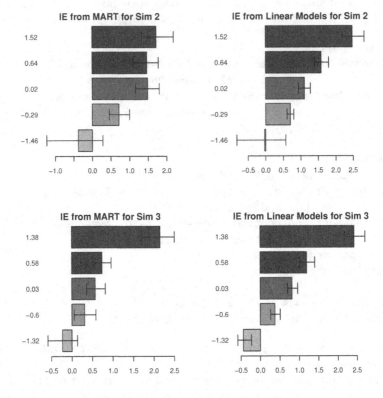

FIGURE 9.3

Indirect moderation effect from Simulations 2 (upper panel) and 3 (lower panel). The y axis is denoted by the mean of the moderator at each quantile. The x-axis is the estimated indirect effect and its confidence interval at each level of the moderator

```
a<-marg.den(x$a.contx$data$dirx[temp.all,xj],
        x$a.contx$data$x[temp.all,vari],
        x$a.contx$data$w[temp.all])
scatter.smooth(a[,1],a[,2],family="gaussian",xlab=colnames(
    x$a.contx$data$dirx)[xj],ylim=xlim,ylab=paste("Mean",
    "m",sep="."),main=paste("mean(mo)","=",
  round(x$a.contx$moder.level$moder.level[l],2)))
  axis(1,at=data$x[temp.all,vari],labels=F)}
dev.off(2)
```

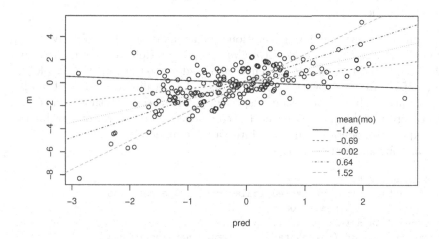

FIGURE 9.4

Exposure-third variable relationship at different level of the moderator for simulation 2.

In the above codes, the function $marg.den(x, y, w)$ is used to order the pair (x, y) by x. The function is included in the appendix of this chapter.

9.2.3 Third-Variable-Moderated TVE

Finally, the third-variable-moderated TVE means that there is a third variable-moderator interaction effect on the outcome. Therefore, the indirect effect of the third-variable would be different at different level of the moderator. The simulation data set is generated as:

$$
\begin{aligned}
c_i, pred_i, mo_i &\overset{ind}{\sim} N(0,1); i = 1, \ldots, n \\
m_i &= pred_i + \epsilon_{1i}; \\
y_i &= pred_i + m_i + c_i + mo_i + mo_i \cdot m_i + \epsilon_{2i}.
\end{aligned}
$$

R Codes to generate the data:

```
#3rd variable-moderated TVE: simulation 3
set.seed(3)
d1=mult.norm(mean1,var1,n)

c1=rnorm(n)
pred=d1[,1]
mo=d1[,2]
```

```
m1=a0+a1*pred+rnorm(n)
y=b0+b1*pred+b2*m1+b3*c1+b4*mo*m1+rnorm(n)
x=cbind(m1,mo,c1)
```

Compared with the data generation scheme in Section 9.2.2, for given mo, the indirect effect of m is also $E(mo_i|mo_i \in q_k) + 1$. However, the moderation effect comes from the interaction effect of m and mo to the outcome. The lower panel of Figure 9.3 shows the indirect effect of m at different levels of mo by MART and by linear regression, respectively. Of the same method and equivalent third-variable indirect effect, the confidence intervals are comparable for the exposure-moderated and the third-variable-moderated TVE.

R Codes for moderation analysis:

```
#nonlinear method
data5<-data.org(x,y,pred=pred,mediator=c(1,3),alpha=0.05,
                alpha2=0.05)
med5=med(data=data5,n=2,nonlinear=T,D=5)
bootmed5<-boot.med(data=data5,n=10,n2=100,nonlinear=TRUE,
                all.model=T,nu=0.001,D=5)
bootmode5=boot.mod(bootmed5,vari="mo",n=10,margin=0.5,
                continuous.resolution=5)
summary(bootmode5,bymed=T)


#linear method
m1=as.matrix(m1)
colnames(m1)="m1"
x2=form.interaction(x,m1,inter.cov="mo")
x2=cbind(x,x2)
data6<-data.org(x2,y,pred=pred,mediator=c(1,3),alpha=0.05,
                alpha2=0.05)
med6=med(data=data6,n=2,xmod="mo")
bootmed6<-boot.med(data=data6,n=10,n2=100,all.model=T,
                xmod="mo")
bootmode6=boot.mod(bootmed6,vari="mo",n=10,
                continuous.resolution=5)
summary(bootmode6)
summary(bootmode6,bymed=T)


#nonlinear method with the interaction term covariate
bootmed5.2<-boot.med(data=data6,n=10,n2=100,all.model=T,
                xmod="mo",nonlinear=T)
bootmode5.2=boot.mod(bootmed5.2,vari="mo",n=10,margin=0.5,
                continuous.resolution=5)
summary(bootmode5.2,bymed=T)


plot2.mma(bootmode5,vari="m1",moderator="mo")
plot2.mma(bootmode6,vari="m1",moderator="mo")
```

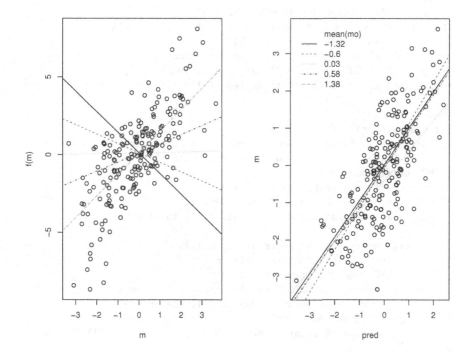

FIGURE 9.5
Exposure-third variable and third variable-outcome relationship at different level of the moderator for Simulation 3.

Figure 9.5 is drawn by the *plot* function of the "mma" package. It shows the pred-m and m-y relationship at each level of the moderator from the linear model. We see that the pred-m relationship is the same but the m-y relationship is different at different level of the moderator.
The R code to draw Figures 9.3 and 9.5:

```
#relationship plot: Figure 9.5
x=bootmode6
full.model=x$a.contx$model$model[[1]]
b1<-predict(full.model,se.fit=T,type=x$a.contx$model$type)$fit
coef<-full.model$coefficients[grep(vari,
                        names(full.model$coefficients))]
#plot the straight line instead of the loess line
location<-"Figure 5.eps"
postscript(location,width=16,height=8)
par(mfrow=c(5,2),mar=c(6,6,1,1),oma=c(3,2,2,4))
```

```
for(q1 in 1:5){
  temp.all=x$a.contx$moder.level$levels[,q1]
  b<-marg.den(x$a.contx$data$x[temp.all,"m1"],b1[temp.all],
         x$a.contx$data$w[temp.all])
  plot(x$a.contx$data$x[,"m1"],b1,type="n",xlab="m",ylab=
     paste("f(","m",")",sep=""),main=paste("mean(mo)=",
 round(x$a.contx$moder.level$moder.level[q1],2)))

  points(b,col=q1)
  if(length(coef)>1){#browser()
    b3=ifelse(is.factor(x$a.contx$moder.level$moder.level),
          1,x$a.contx$moder.level$moder.level[q1])
    b2=coef[2]*b3
    b2.1=coef[2]*b3+coef[1]*mean(b[,1])
    abline(a=mean(b[,2],na.rm=T)-b2.1,b=b2,col=q1)} #
    axis(1,at=x$a.contx$data$x[temp.all,"m1"],labels=F)

    a<-marg.den(x$a.contx$pred.new[temp.all,],
           x$a.contx$data$x[temp.all,"m1"],
           x$a.contx$data$w[temp.all])
    scatter.smooth(a[,1],a[,2],family="gaussian",ylim=xlim,
            xlab=colnames(x$data$dirx)[1],ylab=paste(
            "Mean","m",sep="."),main=paste("mean(mo)=",
            round(x$a.contx$moder.level$moder.level[q1],2)))}
dev.off(2)

#Figure 9.3: Figure of indirect effect
location<-"Figure 3.eps"
postscript(location,width=8,height=8)
par(mfrow=c(2,2),mar=c(6,6,1,1),oma=c(3,2,2,4))
x=summary(bootmode3,bymed=T,plot=F)
re<-sapply(x$a.contx$results$indirect.effect, "[[", 2)
#use the quantile interval
upper<-sapply(x$a.contx$results$indirect.effect, "[[", 6)
lower<-sapply(x$a.contx$results$indirect.effect, "[[", 7)
name1<-c(-1.46,-0.29, 0.02, 0.64, 1.52)
bp <- barplot2(re,horiz=TRUE,main=paste("IE from MART for
        Sim 2"),names.arg=name1,plot.ci=TRUE, ci.u = upper,
        ci.l = lower,cex.names=0.9,beside=FALSE,cex.axis=0.9,
        las=1,xlim=range(c(upper,lower)),col =
 rainbow(length(re), start = 3/6, end = 4/6))
x=summary(bootmode4,bymed=T,plot=F)
re<-sapply(x$a.contx$results$indirect.effect, "[[", 2)
#use the quantile interval
upper<-sapply(x$a.contx$results$indirect.effect, "[[", 6)
lower<-sapply(x$a.contx$results$indirect.effect, "[[", 7)
#name1<-colnames(x$a.contx$results$direct.effect)
bp <- barplot2(re, horiz=TRUE, main=paste("IE from Linear
        Models for Sim 2"),names.arg=name1,plot.ci=TRUE,
```

```
                     ci.u = upper, ci.l = lower,cex.names=0.9,beside=F,
   cex.axis=0.9,las=1,xlim=range(c(upper,lower)),
                     col = rainbow(length(re), start = 3/6, end = 4/6))
x=summary(bootmode5.2,bymed=T,plot=F)
re<-sapply(x$a.contx$results$indirect.effect, "[[", 2)
#use the quantile interval
upper<-sapply(x$a.contx$results$indirect.effect, "[[", 6)
lower<-sapply(x$a.contx$results$indirect.effect, "[[", 7)
name1<-c(-1.32,-0.60, 0.03, 0.58, 1.38)
bp <- barplot2(re, horiz = TRUE, main=paste("IE from MART
                for Sim 3"),names.arg=name1,plot.ci = TRUE,
                ci.u = upper, ci.l = lower,cex.names=0.9,
   beside=FALSE,cex.axis=0.9,las=1,xlim=range(c(
                upper,lower)),col = rainbow(length(re),
   start = 3/6, end = 4/6))
x=summary(bootmode6,bymed=T,plot=F)
re<-sapply(x$a.contx$results$indirect.effect, "[[", 2)
#use the quantile interval
upper<-sapply(x$a.contx$results$indirect.effect, "[[", 6)
lower<-sapply(x$a.contx$results$indirect.effect, "[[", 7)
#name1<-colnames(x$a.contx$results$direct.effect)
bp <- barplot2(re, horiz = TRUE, main=paste("IE from Linear
                Models for Sim 3"),names.arg=name1,plot.ci=T
ci.u = upper, ci.l = lower,cex.names=0.9,
                beside=FALSE,cex.axis=0.9,las=1,xlim=range(c(
upper,lower)),col = rainbow(length(re),
start = 3/6, end = 4/6))
dev.off(2)
```

9.3 Explore the Trend of Racial Disparity in ODX Utilization among Breast Cancer Patients

Breast cancer is the most commonly diagnosed cancer for American women of all races. It is also the second leading cause of cancer death. Breast cancer has been categorized into subgroups for prognosis and treatment purposes. One common way of classifying breast cancer and recommending treatment is based on the expression of estrogen receptor (ER), progesterone receptor (PR), and human epidermal growth factor receptor 2 (HER 2) [7]. The subtype ER positive and/or PR positive (ER+/PR+) and HER2 negative (HER2-) breast cancer is the most common subtype, has the best prognosis and responds well to adjuvant endocrine therapy and/or chemotherapy. However, even within this subtype, patients have different recurrence risk and may respond to chemotherapy differently [54, 55].

Precision medicine has been developed significantly in today's cancer treatment. Oncotype DX® (ODX) is a genomic test which can differentiate

ER+/PR+ and HER2- patients by the risk of recurrence to project prognosis and chemotherapy benefit. ODX test is based on 21-gene expression levels and it produces a recurrence score, a number between 0 and 100. A raised ODX score indicates a higher probability of cancer recurrence and more benefit from chemotherapy. The National Comprehensive Cancer Network (NCCN) cancer treatment guidelines published in 2008 recommended ODX test to patients with ER+/PR+, HER2-, and negative lymph node breast cancer to identify those that are more likely to benefit from chemotherapy. However, research shows that there are racial and ethnic disparities among breast cancer patients in terms of the survival rate, recurrence rate and health-related quality of life [84, 86]. The disparity was also discovered in the use of ODX test [63, 40, 61]. Our previous work has shown that among all female breast cancer patients who were considered to be able to benefit from the ODX exam, non-Hispanic whites had a significantly higher rate of using the test, compared with non-Hispanic blacks [91]. In addition, the proportion of using ODX tests has been increasing over the last decade within both black and white patients. It is interesting to know whether the racial gap in ODX test has been reduced over time during the last decade. Furthermore, if there is a reduction in the gap, what factors contribute to this improvement?

One of the techniques for inferences on the trend of racial disparity in ODX exam is on the moderation effect inferences with a third-variable effect analysis (TVEA) approach. We explore the trend of racial disparities in ODX exam from 2010 to 2015 using a moderation effect inference with TVEA. In this case, the exposure variable is the race (non-Hispanic blacks vs. non-Hispanic whites), the outcome is the use of ODX test (yes/no), and third-variables are all variables that can explain the racial difference in ODX usage. The moderator is the diagnosis year.

9.3.1   Data Description

The Surveillance, Epidemiology, and End Results (SEER) Program is a population-based cancer surveillance program, sponsored by the national cancer institute, which routinely collects standardized demographics information, primary tumor characteristics, first course of treatment, and survival information for all cancer patients through funded SEER cancer registries. It covers approximately 34% of cancer cases of the US population. SEER is an authoritative source for cancer statistics in the US (https://seer.cancer.gov). Genomic Health Inc. (Redwood City, CA) is a unique company that provides the ODX test in the US. SEER program made an effort to link the ODX dataset from the Genomic Health Inc. with SEER data on breast cancer patients diagnosed between 2004 and 2015. More information about the data linkage and available variables can be found at [59].

In this study, we are interested in finding out whether there is a racial disparity in ODX utilization and whether the racial difference changes with time. For this purpose, we select cases from the linked dataset that includes

TABLE 9.1

Racial differences of using the ODX test by year.

Year	Non-hispanic Black	Non-Hispanic White	p-value
2010	34.27%	38.56%	0.0012
2011	37.94%	42.03%	0.0009
2012	38.75%	43.32%	0.0001
2013	43.03%	43.59%	0.6397
2014	44.27%	45.08%	0.4985
2015	43.98%	45.07%	0.3428

non-Hispanic white or black women diagnosed with American Joint Committee on Cancer (AJCC) stage I, II or III, ER+/PR+ and HER2-, and negative lymph node breast cancer. Since SEER began collecting HER2 information from the year 2010, we include cases that were diagnosed between 2010 and 2015 only. In addition, we exclude cases that were identified from death certificate or autopsy only.

Out of 101,104 eligible cases, 14.61% were diagnosed in 2010, 15.89% in 2011, 16.54% in 2012, 17.23% in 2013, 17.51% in 2014 and 18.21% in 2015. A patient is categorized to have an ODX testing if the ODX test was performed within one year after the breast cancer diagnosis. Table 9.1 shows the proportion of the ODX use separated by race and diagnosis year. We found that in general, the proportion of breast cancer patients who had the ODX test increased over time regardless of race and ethnicity. The proportion of ODX testing was higher for NHWs than NHBs in all the years, but the racial differences were diminished and became not significant in 2013 and after.

Other variables used in this study include demographic information for patients (e.g. age and insurance), cancer characteristics (e.g. tumor size and grade), and population characteristics at the county level (e.g. rural/urban and proportion of household below the federal poverty level). Besides race, year of diagnosis and the outcome variable (ODX test), we include 31 variables that can potentially explain the use and the racial disparity in the use of ODX test. In the analysis, we did not include factors on the first-course treatment because ODX test can guide the choice of treatment but not the reverse. Since we are interested in the trend of the racial disparity in ODX and how the effect of contributing factors in explaining the racial disparity varied with year, year-of-diagnosis was used as the moderator. A third-variable is defined as a variable that is significantly related to the exposure variable (race), and significantly related to the outcome (having an ODX test within one year of diagnosis). A covariate is a variable significantly associated with the outcome, but not with the exposure variable. We first tested each variable to identify potential third-variables and covariates. The significance level of tests is set at 0.05. Table 9.2 shows the potential third-variables and covariates as a result of tests.

TABLE 9.2

Racial differences of using the ODX test by year.

Potential Third Variables	
Demographic Variables	age at diagnosis, marital status, insurance
Tumor Characteristics	cancer grade, AJCC stage, tumorsize
County-Level (ACS) Environmental factors	% persons age< 18, % families below poverty, % unemployed, median household income, % < 9th grade, % < high school, % at least bachelors degree, % household > 1 one person/room, % foreign born, % language isolation, % no migration in the year, % move within county, % move within state, % move out-of-state, % move from out-of-US, Rural-Urban code
Potential Covariates	histology type

9.3.2 Third-Variable Effects and the Trend

We use the R package mma to explore the racial disparities in ODX test over the years 2010 to 2015. Since there are potential interaction effects between year and other factors, the third-variable effects need to be estimated at each year. In model fitting, there is no need to include the year × third-variable interaction, MART can find interactions automatically. Figure 9.6 shows the third variable effects of all variables at 2010, ordered from the biggest to the smallest effect. A 95% confidence interval for each estimated third-effect is also shown in the bar plot. The estimated total effect is negative, meaning that compared with non-Hispanic whites, the blacks have a lower probability of using the ODX test. If the third effect is negative as the total effect, it means that if blacks and whites could have the same distribution of the third-variable, the racial disparity in ODX usage can be partially explained by the variable. On the contrary, if a third-effect is positive, it means that if the variable can be manipulated to be equally distributed between blacks and whites, the disparity in ODX would increase, rather than reduce. These variables are called depressors. For example, age is a depressor. For the year 2010, variables % persons age< 18, insurance, % move out-of-state, % no migration in the year, % foreign born, % move within county, and % household > 1 one person/room have significant indirect effects to explain the racial disparity in ODX exam. We explain the direction of some third-variable effects later in this section. After controlling for all other variables, the direct effect of race was still significantly negative. That is, there was remained racial disparity in ODX test that could not be explained by the variables included in this study. The figures for other years are shown in the book website.

To compare the change in racial disparity in ODX usage over years, the *mma* R package provides a summary and plot tool to describe the trend. The left panel of Figure 9.7 describes the direct effect of race with a confidence interval after adjusting for other variables in each year. We find that there was

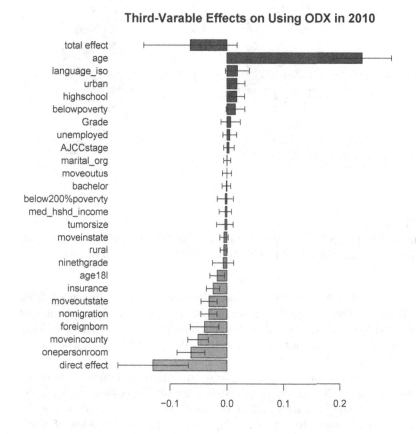

FIGURE 9.6

Third-variable effect of the racial disparity in ODX test in the year 2010.

a significant unexplained racial disparity in ODX test, but the effect decreased from the year 2010 to 2015. The mma package provides a "test.moderation" function, that checks if there is significant direct moderation effect between the predictor (race) and the moderator (year) in explaining the use of ODX. The test with the interaction term in the generalized linear model shows a p-value< 0.001 and the H-statistics is 0.0427. H-statistics is the measure of the significance of the interaction effect in MART. To interpret the H-statistics, readers are referred to Chapter 3 and [27].

The third-variable effects can be similarly described using the plot tool provided by *mma*. An interesting plot is the indirect effect of insurance. We see in the right panel of Figure 9.7 that insurance can significantly explain

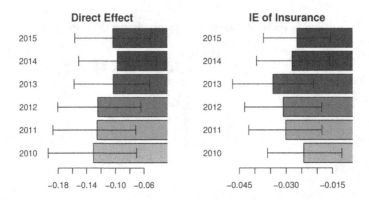

FIGURE 9.7
The unexplained racial disparity (left) and the indirect effect of insurance status on the racial disparity in ODX test over the years 2010 to 2015.

part of the racial disparity. In addition, the indirect effect increased from 2010 to 2013 and then reduced in 2014 and 2015. This might be explained by that from 2012, medicare started to cover a wider range of ODX test. Also, the Obama care act came into force in 2014. We provide plots for all other third-variables in the book website.

Lastly, we use the plot tool to explore how third variables explain the racial disparity in ODX over the years. We use age-at-diagnosis as an example. Figure 9.8 shows the interactive effect of year and age on the odds of using ODX. Looking horizontally over the diagnosis years, the log-odds of using ODX increased with age until around 40, when the use of ODX reached the highest odds. The odds kept high until around the age of 60, when the odds began to decrease with age. MART caught the nonlinear relationship between age and the log-odds of using ODX. Look vertically over the age, the color seems to get lighter over the years, indicating that in general, the odds of using ODX increased over year at the same age. Next, we check if the age at diagnosis is different between blacks and whites. Figure 9.9 displays the age distributions separated by white or black population in the year 2010. It shows that there were a larger proportion of black patients diagnosed between the age 40 and 60, during which patients were more likely to use the ODX diagnosis to guide treatment. Therefore, if the age at diagnosis could be manipulated to distribute similarly between blacks and whites, the racial disparity in ODX would become large rather than be reduced. The age at diagnosis for blacks and whites had the same pattern of distribution for other years and the density plots for each year are provided on the book website. The plots depicting the direction of other third-variable-effect are also included on the book website.

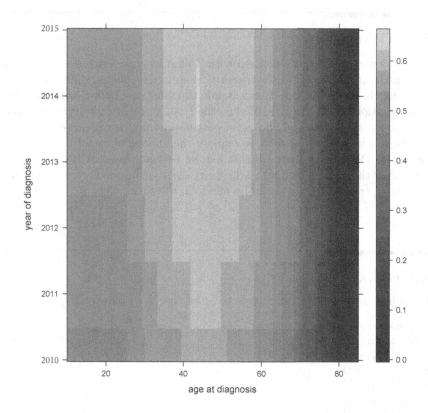

FIGURE 9.8
The interactive effect of year and age on the odds of using ODX.

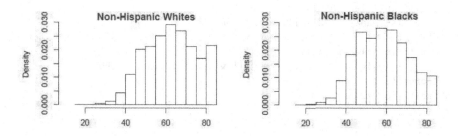

FIGURE 9.9
The age distribution by race in 2010.

The plots can help better understand the third-variable effect and provide explanations on mechanism underlying the racial disparity in ODX test.

9.4   Conclusions

We developed an inference method with TVEA to make inference on interaction/moderation effect. We illustrate the moderation effect at different scenarios using the *mma* package that extends the analysis of third-variable effect to moderation effect. The proposed method can automatically identify significant moderation effect and allows for potential nonlinear relationships among variables. The method is used to explore the change of racial disparities in the use of ODX tests among breast cancer patients from 2010 to 2015. We found that the unexplained racial disparity decreased over the years and some interesting trends of third-variables effect on the racial disparity.

Appendix

The following *marg.den* function is to order the (x, y) pair by x:

```
marg.den<-function(x,y,w=NULL)
{if(!is.null(w))
  w<-w[!is.na(x) & !is.na(y)]
y<-y[!is.na(x)]
x<-x[!is.na(x)]
x<-x[!is.na(y)]
y<-y[!is.na(y)]
z1<-unique(x)
z2<-rep(0,length(z1))
if(is.null(w))
  for (i in 1:length(z1))
    z2[i]<-mean(y[x==z1[i]],na.rm=TRUE)
else
  for (i in 1:length(z1))
    z2[i]<-weighted.mean(y[x==z1[i]],w[x==z1[i]],na.rm=TRUE)
z3<-order(z1)
cbind(z1[z3],z2[z3])
}
```

10

Third-Variable Effect Analysis with Multilevel Additive Models

In research, the experiment data are often collected from different levels. For example, to identify variables that are related with childhood obesity, we consider both environmental (e.g. walkability of the neighborhood) and individual factors (e.g. snacking habit, smoking status and physical activities). When hierarchical databases are considered, third-variable analysis method based on generalized linear models are usually not readily adaptable since the independence assumption among observations is violated. In such cases, hierarchical models, also known as multilevel or mixed-effect models, are more appropriate to fit relationships among variables since these models take into account of dependencies among observations and allow for predictors from different levels of the data ([28]). In this chapter, we discuss a third-variable analysis method based on multilevel models. This chapter is derived in part from an article published in the Journal PLoS One [87] on October 23, 2020, available online at: https://doi.org/10.1371/journal.pone.0241072.

For the hierarchical model, we assume there are two levels of data and refer the individual level as level one or the lower level, and the group level as level two or the higher level. Although more than two levels of hierarchy is possible, this Chapter focuses on two-level databases only.

In third-variable analysis, besides the pathway that directly connect the exposure variable with the outcome, we also explore the $exposure \longrightarrow third- variable \longrightarrow response$ (denoted as $X \to M \to Y$) pathways. When doing third-variable analysis with multilevel models, the level of the variables at the arrow tail should be higher than or equal to that of the arrowhead, since a group level variable may affect an individual level variable but not the reverse ([45]). Therefore, in the 2-level third-variable analysis setting, only the $2 \to 2 \to 2$, $2 \to 2 \to 1$, $2 \to 1 \to 1$, and $1 \to 1 \to 1$ relationships are legible. Multilevel models are necessary to deal with hierarchical database even for the $1 \to 1 \to 1$ relationship. [74] studied the bias brought by using single-level models for estimation when data are hierarchical. [45, 5, 6, 92] proposed third-variable analysis methods for three types of multilevel models. Moreover, [42, 29, 60] proposed alternative methods to test the indirect effects in $2 \to 2 \to 1$, $2 \to 1 \to 1$ relationships. In addition, [89] proposed to use Bayesian third-variable analysis to deal with hierarchical databases.

In this chapter, we introduce the use of generalized additive multilevel models for third-variable analysis with hierarchical data structures. The method is compiled in an R package *mlma*: practitioners can apply the package in research such as health disparity analysis.

This chapter is organized as follows. In Section 10.1, we extend the general definitions of TVEs to the multilevel data settings. We also briefly review the generalized additive multilevel models (GAMM) that are used to model relationships among variables. Based on that, we present the multilevel third-variable analysis with GAMM. In Section 10.2, we illustrate the use of the proposed method in different multilevel data structures and the usage of the *mlma* R package. We then adopt the method in a real data example in Section 10.3: to explore the racial disparity in obesity considering both individual and environmental risk factors. The method can be used with different types of outcomes. For the time-to-event outcome, refers are referred to [93].

10.1 Third-Variable Analysis with Multilevel Additive Models

First, we extend the definitions of third-variable effects within the hierarchical data structure.

10.1.1 Definitions of Third-Variable Effects with Data of Two Levels

The unique data structure in multilevel models raises the potential problem of confounding TVEs from different levels. As pointed out in [92], the relationship between two level-one variables can be decomposed into between-group and within-group components. In particular, the aggregated variables at the second level can be highly related while the relationship may be very weak or even at an opposite direction when considered at the individual level ([62]). For example, [71] points out that the proportion of black residents may be an important variable for the census tract, while it is different from the meaning of ethnicity as an individual-level variable. [2] discussed the difference of the two components extensively. It is important to differentiate the between-group and within-group components in third-variable analysis, where the TVEs can be decomposed to level one and level two effects. To identify the level one and level two TVEs separately, [92] proposed the group-mean centering method (CWC), where they subtracted the group means from individual level variables and added group means as level two covariates. In their paper, [92] showed that the CWC method efficiently separated level one and level two TVEs and resulted in less bias and more power compared with non-centering methods.

Here, we extend the definitions of TVEs with single-level models by [79] to the level one and level two TVEs.

With the generalized definitions of TVEs, [82] has shown that a third-variable analysis can involve multiple exposure variables and multivariate outcomes. The purpose of third-variable analysis is to differentiate the direct effect and indirect effect from each third-variable for each pair of the exposure-outcome relationship. If the outcome is at level two, all exposure and third-variables have to be level two as discussed above. Therefore, a single-level third-variable analysis works. In this chapter, we focus on level one outcomes. If the outcome is at level one, the exposure variable can be a level one or level two variable. The third-variables can be level one or two for a level two exposure but have to be level one for a level one exposure variable.

The conceptual model for TVE of two levels is given in Figure 10.1. In the figure, X denotes the exposure variable, Y is the outcome. There are E_1 level one exposures and E_2 level two exposures. $\mathbf{M}_{ij} = (M_{ij1}, \ldots, M_{ijK})$ is the vector of K potential level one third-variables for the ith object at the jth group. $\mathbf{M}_{.j} = (M_{.j1}, \ldots, M_{.jL})$ is the vector of the L potential level two third-variables or level one third-variables aggregated at level two within group j. $\mathbf{Z}_1$ and $\mathbf{Z}_2$ are other covariates at level one and level two, respectively.

Denote $\mathbf{M}_{ij,-k}$ as the vector $\mathbf{M}_{ij}$ excluding the kth element. Let $M_{ijk}(\mathbf{x}_{ij})$ be a random variable that has a conditional distribution given $\mathbf{X}_{ij} = \mathbf{x}_{ij}$. For an exposure variable X at any level, let u^* be the minimum unit of X, such that if $x \in domain(X)$, then $x + u^* \in domain(X)$. For now, we ignore other covariates $\mathbf{Z}$. Assume effects of exposures and third-variables on the outcome are additive, we have the general definitions of TVEs, following the definitions of TVE from previous chapters, for level one (Definition 10.1) and level two exposure variables (Definition 10.2). Note that a level one exposure can have only level one third-variables while a level two exposure can have both level one and level two third-variables.

Definition 10.1 *For a level one exposure variable X, the level one total effect (TE_1) of X on Y, the level one direct effect $(DE_{1, \backslash M_k})$ of X on Y not from level one third-variable M_k and the level one indirect effect of X on Y through M_k at $X = x_{ij}$ (IE_{1, M_k}) are defined as:*

$$TE_1(x_{ij}) = \lim_{u \to u*} \left[\frac{EY_{ij}(x_{ij} + u, \mathbf{M}_{ij}(x_{ij} + u), x_{.j}, \mathbf{M}_{.j})}{u} \right. \tag{10.1}$$
$$\left. - \frac{EY_{ij}(x_{ij}, \mathbf{M}_{ij}(x_{ij}), x_{.j}, \mathbf{M}_{.j})}{u} \right];$$

$$DE_{1, \backslash k}(x_{ij}) = \lim_{u \to u*} E_{M_{ijk}} \left[\frac{EY_{ij}(x_{ij} + u, \mathbf{M}_{ij,-k}(x_{ij} + u), M_{ijk}, x_{.j}, \mathbf{M}_{.j})}{u} \right.$$
$$\left. - \frac{EY_{ij}(x_{ij}, \mathbf{M}_{ij,-k}(x_{ij}), M_{ijk}, x_{.j}, \mathbf{M}_{.j})}{u} \right]; \tag{10.2}$$

$$IE_{1,k}(x_{ij}) = TE_1(x_{ij}) - DE_{1, \backslash k}(x_{ij}). \tag{10.3}$$

The average level one TVEs are the expected value of the TVEs defined by

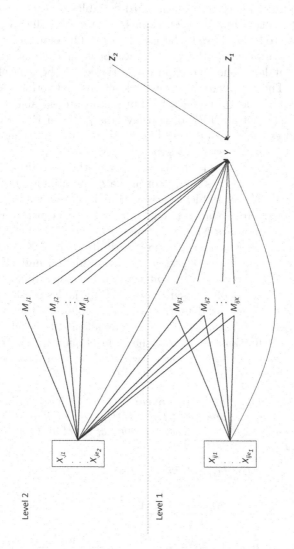

FIGURE 10.1

Conceptual model for two-level third-variable effects.

Definition 10.1: $ATE_1 = E_{ij}[TE_1(x_{ij})]$, $ADE_{1,\backslash k} = E_{ij}[DE_{1,\backslash k}(x_{ij})]$ and $AIE_{1,k} = ATE_1 - ADE_{1,\backslash k}$.

Definition 10.2 *For a level two exposure variable* X, *the level two total effect* *(TE$_2$) of* X *on* Y , *the level two direct effect of* X *on* Y *not from the level one third variable* M_k *(DE$_{21,\backslash k}$), the level two direct effect not from the level two third variable* M_l *(DE$_{22,\backslash l}$), and the level two indirect effect of* X *on* Y *through* M_k *(IE$_{21,k}$) and through* M_l *(IE$_{22,l}$) at* $X = x_{.j}$ *are defined as:*

$$TE_2(x_{.j}) = \lim_{u \to u*} E_i \left[\frac{EY_{ij}(x_{ij}, \mathbf{M}_{ij}(x_{.j} + u), x_{.j} + u, \mathbf{M}_{.j}(x_{.j} + u))}{u} \right.$$
$$\left. - \frac{EY_{ij}(x_{ij}, \mathbf{M}_{ij}(x_{.j}), x_{.j}, \mathbf{M}_{.j}(x_{.j}))}{u} \right]; \tag{10.4}$$

$$DE_{21,\backslash k}(x_{.j}) = \lim_{u \to u*} E_i E_{M_{ijk}} \left[\frac{EY_{ij}(x_{ij}, \mathbf{M}_{ij,-k}(x_{.j}+u), M_{ijk}, x_{.j}+u, \mathbf{M}_{.j}(x_{.j}+u))}{u} \right.$$
$$\left. - \frac{EY_{ij}(x_{ij}, \mathbf{M}_{ij,-k}(x_{.j}), M_{ijk}, x_{.j}, \mathbf{M}_{.j}(x_{.j}))}{u} \right]; \tag{10.5}$$

$$DE_{22,\backslash l}(x_{.j}) = \lim_{u \to u*} E_i E_{M_{.jl}} \left[\frac{EY_{ij}(x_{ij}, \mathbf{M}_{ij}(x_{.j}+u), x_{.j}+u, \mathbf{M}_{.j,-l}(x_{.j}+u), M_{jl})}{u} \right.$$
$$\left. - \frac{EY_{ij}(x_{ij}, \mathbf{M}_{ij}(x_{.j}), x_{.j}, \mathbf{M}_{.j,-l}(x_{.j}), M_{jl})}{u} \right]; \tag{10.6}$$

$$IE_{21,k}(x_{.j}) = TE_2(x_{.j}) - DE_{21,\backslash k}(x_{.j}); \tag{10.7}$$

$$IE_{22,l}(x_{.j}) = TE_2(x_{.j}) - DE_{22,\backslash l}(x_{.j}). \tag{10.8}$$

The average level two TVEs are the TVEs defined by Definition 10.2 averaged at the group level: $ATE_2 = E_j[TE_{2,j}(x_{.j})]$, $AIE_{21,k} = E_j[IE_{21,jk}(x_{.j})]$ and $AIE_{22,l} = E_j[IE_{22,jl}(x_{.j})]$.

10.1.2   Multilevel Additive Models

We use multilevel additive models to build relationships among variables of hierarchical structure. An additive model is a nonlinear regression method that was first proposed by [25]. A multilevel additive model can deal with both nonlinear covariate effects and cluster-specific heterogeneity ([46]). It is now gaining rapid popularity in psychological and social research ([11]). Using the notations in Section 10.1.1, assume that we have L level two and K level one third-variables. In addition, assume that there are E_1 level one and E_2 level two exposure variables, denoted as $\mathbf{X}_{ij}^T = \{X_{ij1}, \ldots, X_{ijE_1}\}$ and $\mathbf{X}_{.j}^T = \{X_{.j1}, \ldots, X_{.jE_2}\}$, respectively. We propose the following linear additive multilevel models for multilevel third-variable analysis. The boldfaced letter indicates a vector of functions or numbers.
For level two third-variables, $M_{.jl}$, $l = 1, \ldots, L$:

$$g_{.l}(E(M_{.jl})) = \alpha_{0l} + \sum_{e=1}^{E_2} \boldsymbol{\alpha}_{2le}{}^T \mathbf{f}_{2le}(X_{.je}).$$

For level one third-variables, M_{ijk}, $k = 1, \ldots, K$:

$$g_{1k}(E(M_{ijk})) = u_{0jk} + \sum_{e=1}^{E_1} \boldsymbol{\alpha}_{1ke1}{}^T \mathbf{f}_{1ke1}(X_{ije})$$

$$+ \sum_{e=1}^{E_2} \boldsymbol{\alpha}_{2ke1}{}^T \mathbf{f}_{2ke1}(X_{.je});$$

$$u_{0jk} = \alpha_{00k} + r_{0jk}.$$

The full model:

$$E(Y_{ij}) = u_{0j} + \sum_{e=1}^{E_1} \boldsymbol{\beta}_{1e}{}^T \mathbf{f}_{1e}(X_{ije}) + \sum_{e=1}^{E_2} \boldsymbol{\beta}_{2e}{}^T \mathbf{f}_{2e}(X_{.je})$$

$$+ \sum_{k=1}^{K} \boldsymbol{\beta}_{3k}{}^T \mathbf{f}_{3k}(M_{ijk}) + \sum_{l=1}^{L} \boldsymbol{\beta}_{4l}{}^T \mathbf{f}_{4l}(M_{.jl});$$

$$u_{0j} = \beta_{00} + r_{0j}.$$

In the models, r_{0jk} and r_{0j} are second-level random errors with mean zero and finite variances. $\mathbf{f}(\cdot)$ is a vector of function/transformation on $\cdot$. The transformation enables modeling nonlinear relationships among variables. We assume that all the transformation functions are first-order differentiable. $\boldsymbol{\alpha}$ and $\boldsymbol{\beta}$ are coefficient vectors for transformed variables in predicting the response variable on the left side of each equation. In addition, $g(\cdot)$s are the link functions that link the expected response variable with the right-hand-side of each equation, the systematic component of a generalized linear model. For example, a binary M_{ijk} may have a link function $g_{1k} = logit(P(M_{ijk} = 1))$. Similarly, a link function can be used on the outcome. With the link function, we can deal with different types of third-variables and outcomes.

10.1.3 Third-Variable Effects with Multilevel Additive Model

Based on the definitions of TVEs, we derive the TVEs in Theorems 10.1 and 10.2. In the theorems, $f'(x)$ denotes the first derivative of function f on the random variable X and realized at $X = x$. In addition, g^{-1} denotes the inverse function of g. We further denote $\mu_{ijk} = E(M_{ijk})$ and $\mu_{.jk} = E(M_{.jk})$. The proofs of theorems are given after each theorem.

Theorem 10.1 *With the relationships among variables built by Section 10.1.2, the TVEs for level one exposure variable $X_{ije}, e = 1, \ldots, E1$ on the outcome variable Y are:*

$$IE_{1,k}(x_{ije}) = [\boldsymbol{\alpha}_{1ke1}{}^T \mathbf{f}'_{1ke1}(x_{ije}) \cdot g_{1k}^{-1'}(\mu_{ijk})] \times [\boldsymbol{\beta}_{3k}{}^T \mathbf{f}_{3k}'(m_{ijk})],$$

$$k = 1, \ldots K$$

$$DE_1(x_{ije}) = \boldsymbol{\beta}_{1e}{}^T \mathbf{f}_{1e}{}'(x_{ije})$$

$$TE_1(x_{ije}) = DE_1(x_{ije}) + \sum_{k=1}^{K} IE_{1,k}(x_{ije})$$

Proof *(Proof for Theorem 10.1) For the third-variable effects at the first level, by the multilevel additive models described in Section 10.1.2, we have that*

$$
\begin{aligned}
E(Y_{ij}) &= u_{0j} + \sum_{e=1}^{E_1} \boldsymbol{\beta}_{1e}{}^T \mathbf{f}_{1e}(X_{ije}) + \sum_{e=1}^{E_2} \boldsymbol{\beta}_{2e}{}^T \mathbf{f}_{2e}(X_{.je}) \\
&\quad + \sum_{k=1}^{K} \boldsymbol{\beta}_{3k}{}^T \mathbf{f}_{3k}(M_{ijk}) + \sum_{l=1}^{L} \boldsymbol{\beta}_{4l}{}^T \mathbf{f}_{4l}(M_{.jl}).
\end{aligned}
$$

In addition, we have that for $k = 1, \ldots, K$,

$$g_{1k}(E(M_{ijk})) = u_{0jk} + \sum_{e=1}^{E_1} \boldsymbol{\alpha}_{1ke1}{}^T \mathbf{f}_{1ke1}(X_{ije}) + \sum_{e=1}^{E_2} \boldsymbol{\alpha}_{2ke1}{}^T \mathbf{f}_{2ke1}(X_{.je}).$$

Denoting that $\mu_{ijk} = E(M_{ijk})$, by Equation (1) in Definition 1, we have

$$
\begin{aligned}
TE_1(x_{ije}) &= \left. \frac{\partial E(Y_{ije})}{\partial X_{ije}} \right|_{X_{ije}=x_{ije}} \\
&= \left[\frac{\partial \boldsymbol{\beta}_{1e}^T \mathbf{f}_{1e}(X_{ije})}{\partial X_{ije}} + \sum_{k=1}^{K} \frac{\partial \boldsymbol{\beta}_{3k}^T \mathbf{f}_{3k}(M_{ijk})}{\partial M_{ijk}} \cdot \frac{\partial M_{ijk}}{\partial X_{ije}} \right]_{X_{ije}=x_{ije}} \\
&= \boldsymbol{\beta}_{1e}{}^T \left. \frac{\partial \mathbf{f}_{1e}(X_{ije})}{\partial X_{ije}} \right|_{X_{ije}=x_{ije}} + \sum_{k=1}^{K} \left[\boldsymbol{\beta}_{3k}^T \left. \frac{\partial \mathbf{f}_{3k}(M_{ijk})}{\partial M_{ijk}} \right|_{M_{ijk}=m_{ijk}} \right. \\
&\quad \left. \cdot \boldsymbol{\alpha}_{1ke1}^T \left. \frac{\partial \mathbf{f}_{1ke1}(X_{ije})}{\partial X_{ije}} \right|_{X_{ije}=x_{ije}} \cdot \left. \frac{\partial g_{1k}^{-1}(M_{ijk})}{\partial M_{ijk}} \right|_{M_{ijk}=\mu_{ijk}} \right] \\
&= DE_1(x_{ije}) + \sum_{k=1}^{K} IE_{1,k}(x_{ije})
\end{aligned}
$$

Theorem 10.2 *With the relationships among variables built by Section 10.1.2, the TVEs for level two exposure variable $X_{.je}, e = 1, \ldots, E2$ on the outcome variable Y are:*

$$IE_{22,l}(x_{.je}) = [\boldsymbol{\alpha}_{2le}{}^T \mathbf{f}_{2le}'(x_{.je}) \cdot g_{.l}^{-1}{}'(\mu_{.jl})] \times [\boldsymbol{\beta}_{4l}{}^T \mathbf{f}_{4l}'(m_{.jl})], l = 1, \ldots L$$

$$IE_{21,k}(x_{.je}) = [\boldsymbol{\alpha}_{2ke1}{}^T \mathbf{f}_{2ke1}'(x_{.je}) \cdot g_{1k}^{-1}{}'(\mu_{ijk})] \times [\boldsymbol{\beta}_{3k}{}^T E\mathbf{f}_{3k}'(m_{ijk})], k = 1, \ldots K$$

$$DE_2(x_{.je}) = \boldsymbol{\beta}_{2e}{}^T \mathbf{f}_{2e}'(x_{.je})$$

$$TE_1(x_{ije}) = DE_2(x_{.je}) + \sum_{k=1}^{K} IE_{21,k}(x_{.je}) + \sum_{l=1}^{L} IE_{22,l}(x_{.je})$$

Proof *(Proof for Theorem 10.2) For the third-variable effects at the second level, by the multilevel additive models described in Section 10.1.2, we have that*

$$
E(Y_{ij}) = u_{0j} + \sum_{e=1}^{E_1} {\beta_{1e}}^T \mathbf{f}_{1e}(X_{ije}) + \sum_{e=1}^{E_2} {\beta_{2e}}^T \mathbf{f}_{2e}(X_{\cdot je})
$$

$$
+ \sum_{k=1}^{K} {\beta_{3k}}^T \mathbf{f}_{3k}(M_{ijk}) + \sum_{l=1}^{L} {\beta_{4l}}^T \mathbf{f}_{4l}(M_{\cdot jl}).
$$

In addition, we have that for $k = 1, \ldots, K$,

$$
g_{1k}(E(M_{ijk})) = u_{0jk} + \sum_{e=1}^{E_1} {\alpha_{1ke1}}^T \mathbf{f}_{1ke1}(X_{ije}) + \sum_{e=1}^{E_2} {\alpha_{2ke1}}^T \mathbf{f}_{2ke1}(X_{\cdot je});
$$

and for $l = 1, \ldots, L$,

$$
g_{\cdot l}(E(M_{\cdot jl})) = \alpha_{0l} + \sum_{e=1}^{E_2} {\alpha_{2le}}^T \mathbf{f}_{2le}(X_{\cdot je}).
$$

$$
TE_2(x_{\cdot je}) = \left. \frac{\partial E(Y_{ije})}{\partial X_{\cdot je}} \right|_{X_{\cdot je}=x_{\cdot je}}
$$

$$
= \left[\frac{\partial \beta_{2e}^T \mathbf{f}_{2e}(X_{\cdot je})}{\partial X_{\cdot je}} + \sum_{k=1}^{K} \frac{\partial \beta_{3k}^T \mathbf{f}_{3k}(M_{ijk})}{\partial M_{ijk}} \cdot \frac{\partial M_{ijk}}{\partial X_{\cdot je}} \right.
$$

$$
\left. + \sum_{l=1}^{L} \frac{\partial \beta_{4l}^T \mathbf{f}_{4l}(M_{\cdot jl})}{\partial M_{\cdot jl}} \cdot \frac{\partial M_{\cdot jl}}{\partial X_{\cdot je}} \right]_{X_{\cdot je}=x_{\cdot je}}
$$

$$
= \left. \beta_{2e}^T \frac{\partial \mathbf{f}_{2e}(X_{\cdot je})}{\partial X_{\cdot je}} \right|_{X_{\cdot je}=x_{\cdot je}} + \sum_{k=1}^{K} \left[\beta_{3k}^T E \left[\left. \frac{\partial \mathbf{f}_{3k}(M_{ijk})}{\partial M_{ijk}} \right|_{M_{ijk}=m_{ijk}} \right] \right.
$$

$$
\left. \times \alpha_{2ke1}^T \left. \frac{\partial \mathbf{f}_{2ke1}(X_{\cdot je})}{\partial X_{\cdot je}} \right|_{X_{\cdot je}=x_{\cdot je}} \cdot \left. \frac{\partial g_{1k}^{-1}(M_{ijk})}{\partial M_{ijk}} \right|_{M_{ijk}=\mu_{ijk}} \right]
$$

$$
+ \sum_{l=1}^{L} \left[\beta_{4l}^T \left. \frac{\partial \mathbf{f}_{4l}(M_{\cdot jl})}{\partial M_{\cdot jl}} \right|_{M_{\cdot jl}=m_{\cdot jl}} \cdot \alpha_{2le}^T \left. \frac{\partial \mathbf{f}_{2le}(X_{\cdot je})}{\partial X_{\cdot je}} \right|_{X_{\cdot je}=x_{\cdot je}} \right.
$$

$$
\left. \times \left. \frac{\partial g_{\cdot l}^{-1}(M_{\cdot jl})}{\partial M_{\cdot jl}} \right|_{M_{\cdot jl}=\mu_{\cdot jl}} \right]
$$

$$
= DE_2(x_{\cdot je}) + \sum_{k=1}^{K} IE_{21,k}(x_{\cdot je}) + \sum_{l=1}^{L} IE_{22,l}(x_{\cdot je})
$$

10.1.4   Bootstrap Method for Third-Variable Effect Inferences

Finally, we use the bootstrap method to calculate the variances of the TVEs. In particular, at the group level, a bootstrap sample of the same size for each group is drawn with replacement from the original data set. Then multilevel additive models are fitted based on the bootstrap sample to get the estimates of βs and αs, based on which the TVEs can be calculated by Theorems 10.1 and 10.2. The above process repeats many times and inferences can be made based on the repeated estimates. The bootstrap method is adopted in the R package *mlma* to estimate the variances of estimates and to build up confidence intervals. The *mlma* package is available from the Comprehensive R Archive Network (CRAN) at https://cran.r-project.org/web/packages/mlma/index.html and illustrated in the following Sections.

10.2   The *mlma* R Package

The authors developed a R package, *mlma*, for multilevel third-variable analysis. The analysis is based on multilevel additive models where nonlinear transformations of variables are allowed. The package **mlma** contains three groups of functions: function *data.org* is used to prepare data sets for analysis – transforming variables, dichotomizing categorical third-variables, and getting the derivatives of the transformation functions. The functions *mlma* and *boot.mlma* are used for statistical inferences on the TVEs. The former estimates the TVEs and the latter generates bootstrap samples from the original data sets and does the third-variable analysis based on the bootstrap samples. The estimates of TVEs from the bootstrap samples are then used for statistical inferences. The third group of functions is generic functions – *print*, *summary* and *plot* on the objects generated by the *mlma* and *boot.mlma* functions. In this section, we exemplify the use of the package. The results of the following sections are all generated from the package.

To use the R package *mlma*, we first install the package in R and load it.

```
install.packages("mlma")
library(mlma)
```

10.2.1   A Simulated Dataset

To illustrate the use of the package, we generate a dataset with two levels. In the simulation, there are one level one exposure that is binary and one level two exposure that is continuous. There are also two third-variables, one at each level. The level one third-variable is continuous while the level two third-variable is binary. The variables are generated by the following code:

```
set.seed(1)
n=20        # the number of observations in each group
J<-600/n    # there are 30 groups
```

```
level=rep(1:J,each=n)
alpha_211<-0.8        #covariates coefficients
alpha_1111<-0.8
alpha_2111<-0.8
beta_1<-0.4
beta_2<-0.4
beta_3<-0.4
beta_4<-0.4
beta_5<-0.4
v1=5                  #the level 1 variance
v2=v1/5               #the level 2 variance


#The exposure variables
x1<-rbinom(600,1,0.5) #binary level 1 exposure, xij
x2<-rep(rnorm(J),each=n) #continuous level 2 exposure


#The third-variables
m2<-rep(rbinom(J,1,exp(alpha_211*unique(x2^2))/(1+exp(alpha_211*
          unique(x2^2)))),each=n)     #level 2 binary third-variable
u1<-rep(rnorm(J,0,0.5),each=n) #level 2 variance for mij
e1<-rnorm(n*J)   #level 1 variance for mij
m1<-u1+alpha_1111*x1+alpha_2111*x2+e1 #level 1 continuous third-variable


#The response variable
u0<-rep(rnorm(J,0,v2),each=n)
e0<-rnorm(n*J,0,v1)
y<-u0+beta_1*x1+beta_2*x2+beta_3*ifelse(x2<=0,0,log(1+x2))
   +beta_4*m1+beta_5*m2+e0
```

In summary, we generate the following variables:

- A level-one binary exposure variable, $x1$ or x_{ij};

- A level-two continuous exposure variable, $x2$ or $x_{.j}$;

- A level-one continuous third-variable, $m1$ or m_{ij};

- A level-two binary third-variable, $m2$ or $m_{.j}$;

- A level-one continuous outcome y or y_{ij}.

- There are n observations each group, $i = 1, \ldots, n$, and $600/n$ groups, $j = 1, \ldots, 600/n$.

The relationship among variables are that:

$$
\begin{aligned}
logit(m_{.j} = 1) &= 0.8x_{.j}^2, \\
m_{ij} &= u_{0j} + 0.8x_{.j} + 0.8x_{.j} + \epsilon_{0ij}, \\
y_{ij} &= u_{1j} + 0.4x_{ij} + 0.4x_{.j} + 0.4I(x_{.j} > 0)log(x_{.j} + 1) \\
&\quad + 0.4m_{ij} + 0.4m_{.j} + \epsilon_{1ij},
\end{aligned}
$$

where $u_{0j} \sim N(0, 0.5), \epsilon_{0ij} \sim N(0,1), u_{1j} \sim N(0, v_2)$ and $\epsilon_{1ij} \sim N(0, v_1)$ are independently generated random errors at both levels. The level two random error for the response variable is set as one-fifth of the level one variance – $v_2 = v_1/5$, which makes the intra-class correlation (ICC) .17. As pointed out by [12], this medium ICC value facilitates model convergence. v_2 is chosen to be 5.

10.2.2 Data Transformation and Organization

The *data.org* function is used to do the transformations for variables before the third-variable analysis. To run the function, the minimum inputs are the exposure variable(s) (x) and the third-variables(s) (m). The response variable (y) is required only when its level (*levely*) is not given. The argument *levelx* is to identify the levels of the exposure variable. *levelx* does not need to be provided. The function can automatically decide the level of the exposure variable(s). If any of the exposure variable is binary or categorical, *xref* is used to specify the reference group of the exposure variable.

The arguments $l1$ and $l2$ specify the column numbers in m the continuous third-variables at level one or level two, respectively. $c1$ and $c2$ refer to the categorical third-variables where $c1r$ and $c2r$ specify the reference group, respectively. $l1, l2, c1$ and $c2$ do not have to be provided. If not provided, the function *data.org* checks each column of m and decides whether each variable belongs to level 1 or 2, and be continuous or categorical.

level is a vector that identifies the group number for each observation. If not provided, *level* will assign each observation to a different group. That is, assume no observation is nested within the same group of any other observations. *weight* defines the weight of each observation.

$f01y$ and $f10y$ specify the desired transformation of exposures at level two or level one, respectively in explaining the response variable. $f01y$ and $f10y$ are lists. The first item of $f01y$ or $f10y$ identifies the column number(s) of the exposure variable in x that needs to be transformed, and then in that order, each of the rest items list the transformation functional expressions for each exposure. For example, `f10y=list(1,c("x^2","log(x)"))` means that column 1 of x is a level 1 exposure. It needs to be transformed to its square form and natural log form to predict the response variable. If not specified in $f01y$ or $f10y$, the exposure variable in x keeps its original format without any transformation. In our simulation data, the level two exposure, $x[,2]$, is transformed to itself, $x_{\cdot j}$, and $I(x_{\cdot j} > 0) \times log(x_{\cdot j} + 1)$. Therefore, we define `f01y=list(2,c("x","ifelse(x>0,log(x+1),0)"))`. Similarly, $f02ky$ and $f20ky$ defines the transformation of level two and level one third-variables (in m), respectively in explaining y. In this example, all third-variables are in their original formats to explain y, so $f02ky$ and $f20ky$ are not set up.

$f01km1$ and $f01km2$ are arguments that define transformation of level two exposures in explaining level one or level two third-variable(s), respectively. Since only higher or equivalent level variables can be used as predictors, level one exposures can only be predictors for level one third-variables. $f10km$ defines the transformation of level one exposure(s) in explaining level one third-variable(s). $f01km1$, $f01km2$ and $f10km$ are lists, the first item of which is a matrix of two columns. The first column indicates the column number of the third-variable in m to be explained. The second column indicates the column number of the exposure in x to be transformed to explain the third-variable identified by the first column. By the order of the rows of the matrix, each of the rest list items of $f01km1$, $f01km2$ or $f10km$ list the transformation functional expressions for the exposure (identified by column 2 of the matrix) in explaining each third-variable (identified by column 1). In our example, level two third-variable $m[, 2]$ is explained by the level two exposure $x[, 2]$ in the form of $x2^2$. Therefore, we set the argument f01km2=list(matrix(c(2,2),1,2),"x\^2").

Note that if there are level two third-variables but no level two exposure variable is defined, the level one exposure variable(s) will be aggregated at level two to form the level two exposure variable(s).

The following codes prepare for the data and perform the transformations for the simulation data set generated in Section 10.2.1. Note that the transformation functions can be set in different ways. Besides those in the example, we can also use the natural spline bases (e.g. ns(x,df=5)) and piecewise functions (e.g. ifelse(x<0,0,sqrt(x))).

```
example1<-data.org(x=cbind(x1=x1,x2=x2), m=cbind(m1=m1,
              m2=m2),f01y=list(2,c("x","ifelse(x>0,
              log(x+1),0)")),level=level,f01km1=
              list(matrix(c(1,2),1,2),"x^2"))
```

10.2.3   Multilevel Third-Variable Analysis

10.2.3.1   The *mlma* Function

The function *mlma* can be executed based on the results from *data.org* or on the original arguments of *data.org*. In addition, the response variable needs to be set up by y. If the response variable is categorical, $yref$ is used to specify the reference group. The *random* argument is to set up the random effect part for the response variable and *random.m*1 is for the third-variables.

The argument *covariates* includes the data frame of all covariates for the response variable and/or third-variables. For the response variable, covariates are defined as those variables used to explain y, but are not related or caused by the exposure variable(s). Arguments $cy1$ and $cy2$ specify the column numbers of level one and two covariates respectively in *covariates* for the response variable. *cm* specifies the covariates for third-variables.

If the joint effect of a group of third-variables is of interest, the group can be set up with the *joint* argument. Finally, if users are interested in the third-variable effects on a new set of exposure and third-variables, the new sets can also be set. Please read the help menu of the package.

The following codes perform the multilevel third-variable analysis and report results:

```
mlma.e1<-mlma(y=y,data1=example1,intercept=F)
mlma.e1
Level 2 Third Variable Effects:
     TE   DE m2.1   m1
x2 1.05 0.62 0.01 0.43
Level 1 Third Variable Effects:
     TE  DE   m1
x1 1.04 0.5 0.54
```

The result of third-variable effect analysis shows the third-variable effect from different levels. The direct effect, indirect effects and total effect are shown for each exposure-response pair of variables. For the above example, the level one total effect from $x1$ to y is 1.04, of which direct effect is 0.5 and the indirect effect from $m1$ is 0.54. The level two total effect between $x2$ and y is 1.05, in which the direct effect is 0.62, the indirect effect from $m2$ is 0.01, and from $m1$ is 0.43.

10.2.3.2 The *Summary* Function for Multilevel Third-Variable Analysis

To understand the direction of third-variable effects, the *summary* function for the mlma object (output from the mlma function) provides the ANOVA type III tests of the exposure variables and third-variables in the full model to estimate the response variable. It also provides the ANOVA tests of the exposure variable(s) in predicting each third-variable. Using the results, users can decide which variables should be included as third-variables and which ones should be used as covariates, and then rerun the multilevel third-variable analysis. The following are results by running summary(mlma.e1).

```
1. Anova on the Full Model:
Analysis of Deviance Table (Type III Wald chisquare tests)

Response: y
      Chisq Df Pr(>Chisq)
x1    1.3500 1    0.245285
x2.1 0.3080 1    0.578897
x2.2 0.0998 1    0.752076
m2.2 0.0403 1    0.840988
m1    7.5740 1    0.005922 **
---
Signif. codes:  0 '***' 0.001 '**' 0.01 '*' 0.05 '.' 0.1 ' ' 1

2. Anova on models for Level 1 mediators:
$m1
Analysis of Deviance Table (Type III Wald chisquare tests)
```

```
Response: y
    Chisq Df Pr(>Chisq)
x2  51.817  1  6.092e-13 ***
x1 111.038  1  < 2.2e-16 ***
---
Signif. codes:  0 '***' 0.001 '**' 0.01 '*' 0.05 '.' 0.1 ' ' 1

3. Anova on models for Level 2 mediators:
$m2
Analysis of Deviance Table (Type III tests)

Response: y
      LR Chisq Df Pr(>Chisq)
x2.2.1   9.5608  1   0.001988 **
---
Signif. codes:  0 '***' 0.001 '**' 0.01 '*' 0.05 '.' 0.1 ' ' 1
```

The first part of the above result gives the importance of each variable in predicting the outcome. We found that only $m1$ is significant in predicting y. This could be due to the small coefficients (0.4) and multi-collinearity among predictors for y. The second and third parts give the ANOVA results when using exposure(s) and covariate(s) to predict level one and two third-variables respectively.

Finally, if the fitted models are of interests, users can find them through the following codes:

```
> mlma.e1$f1   #the full model
Linear mixed model fit by REML ['lmerMod']
Formula: y ~ x1 + x2.1 + x2.2 + m2.2 + m1 - 1 + (1 | level)
   Data: data.frame(temp.data)
REML criterion at convergence: 3660.149
Random effects:
 Groups   Name        Std.Dev.
 level    (Intercept) 1.182
 Residual             5.040
Number of obs: 600, groups:  level, 30
Fixed Effects:
    x1     x2.1    x2.2     m2.2      m1
0.5021  0.4120  0.5781  0.1328  0.4979

> mlma.e1$fm1  #models for level 1 third-variables
[[1]]  #This is the order of level 1 third-variables in m
[1] 1

[[2]]  #For the third-variable m[,1]
Linear mixed model fit by REML ['lmerMod']
Formula: y ~ x2 + x1 - 1 + (1 | level)
   Data: data.frame(temp.data)
```

```
REML criterion at convergence: 1863.169
Random effects:
 Groups    Name         Std.Dev.
 level     (Intercept)  0.5675
 Residual               1.0878
Number of obs: 600, groups:  level, 30
Fixed Effects:
     x2      x1
 0.8537  0.8955


> mlma.e1$fm2  #models for level 2 third-variables $
[[1]]  #This is the order of level 2 third-variables in m
[1] 2

[[2]]  #The fitted model for m[,2]

Call:  glm(formula = frml.m, family = binomial(link = "logit"),
    data = data.frame(temp.data))

Coefficients:
x2.2.1
 1.277


Degrees of Freedom: 30 Total (i.e. Null);  29 Residual
Null Deviance:     41.59
Residual Deviance: 32.03  AIC: 34.03
```

10.2.3.3 The Plot Function for the mlma Object

The *plot.mlma* function helps depict the directions of third-variable effects based on the mlma object. Without specifying the third-variable (by *var*), the *plot* function plots the estimated relative third-variable effects (third-variable effect/total effect). Figure 10.2 is drawn by the code plot(mlma.e1).

If a variable is specified, the *plot* function draws the indirect effect of the third-variable, and its marginal relationship with the response variable and with the exposure variable respectively at each level (for level one third-variable) or at the second level (for level two third-variables).

Figure 10.3 is the results from plot(mlma.e1,var="m1"). The left panel of Figure 10.3 shows the level one third-variable effect from $x1$ to y. Since the exposure variable $x1$ is binary, there is only one indirect effect of $x1$ at its reference level. The upper figure shows the boxplots of $m1$ at different levels of $x1$. We can see that in general, $m1$ is higher at $x1 = 1$, compared to $x1 = 0$. The lower plot shows the actual (dots) and fitted relationship (the line) between $m1$ and y. We can see a rough increasing relationship between $m1$ and y. The right panel shows the indirect effect from the level two exposure $x2$ to $m1$ and then to y. The upper figure shows the indirect effect of $m1$ at different value of $x2$, we see that the indirect effect is roughly constant. The middle figure shows the differential effect between $x2$ and m. The differential effect

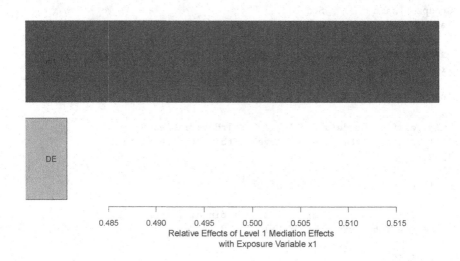

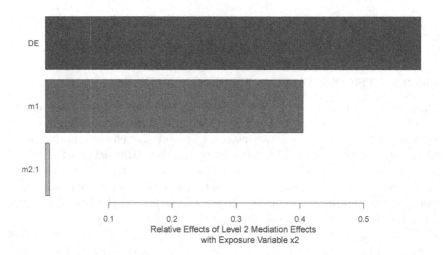

FIGURE 10.2
The relative effects at both levels by *plot(mlma.e1)*.

is the changing rate of $m1$ with $x2$, which is a constant at around 0.4. This indicates a linear relationship between $m1$ and $x2$. Finally, the lower panel shows the relationship between $m1$ and y aggregated at level two, which is a roughly increasing relationship.

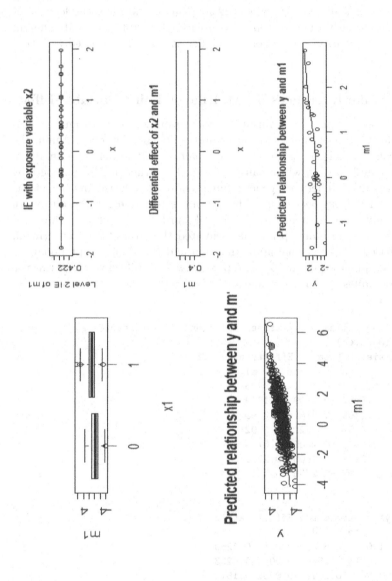

FIGURE 10.3

The figures by *plot(mlma.e1,var="m1")*. The left panel is for the level one $x1 - y$ relationship and the right panel is for the level two $x2 - y$ relationship.

Figure 10.4 shows the result from the code
`plot(mlma.e1,var="m2")`. The upper plot is the indirect effect (IE) of $m2$
at different value of $x2$. We see a quadratic relationship. The middle graph
is the differential effect between $x2$ and $m2$. The changing rate of $m2$ with
$x2$ is increasing, indicating a quadratic relationship between $m2$ and $x2$ with
a positive coefficient for $m2^2$. Since $m2$ is a binary variable, the lower plot
shows the relationship between $m2$ and y through boxplots of y at different
levels of $m2$. In general, the outcome is higher at $m2 = 1$ compared with that
at $m2 = 0$.

10.2.4   Make Inferences on Multilevel Third-Variable Effect

Finally, the *boot.mlma* function uses the bootstrap method to estimate third-
variable effects, the estimated variances and confidence intervals. Again, the
analysis can be built on the results from *data.org*. The default number of
bootstrap sample is 100, which can be changed to other desired numbers by
the argument *boot*. There are generic functions to summarize and depict the
object returned from *boot.mlma*. The *summary* function for the output of
boot.mlma gives the inference results for all third-variable effects. Two confidence intervals are built up for the estimated third-variable effects. (lwbd,
upbd) is based on the normal approximation and (lwbd.quan, upbd.quan) is
built by the quantiles of the bootstrap results. The following codes run the
bootstrap samples for third-variable analysis and then the summarized results
are shown after.

```
> boot.e1<-boot.mlma(y=y,data1=example1,echo=F,intercept = F)
> summary(boot.e1)
MLMA Analysis: Estimated Effects at level 1:
                te        de        m1
est          1.0447    0.5021    0.5426
mean         0.5937    0.4909    0.1029
sd           0.3709    0.3782    0.0899
upbd         1.3207    1.2322    0.2792
lwbd        -0.1332   -0.2505   -0.0734
upbd.quan    1.2681    1.1508    0.2791
lwbd.quan   -0.1997   -0.2705   -0.0258


MLMA Analysis: Estimated Effects at level 2:
                te       de       m2.1       m1
est          1.0496   0.6183    0.0063   0.4250
mean         1.0871   0.6574    0.0045   0.4253
sd           0.2091   0.2847    0.0202   0.1518
upbd         1.4970   1.2155    0.0441   0.7228
lwbd         0.6773   0.0993   -0.0352   0.1277
upbd.quan    1.4675   1.2840    0.0431   0.6923
lwbd.quan    0.6869   0.1685   -0.0368   0.1585
```

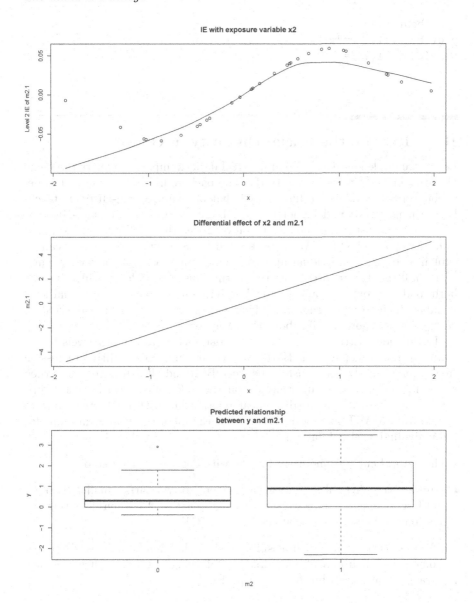

FIGURE 10.4
The figures by *plot(mlma.e1,var="m2")*.

The *plot* function for the *boot.mlma* objects works similarly for that of the *mlma* objects but confidence intervals for all estimations are added. For the above third-variable analysis, reader can use the following codes to delineate relationships of interest.

```
plot(boot.e1)
plot(boot.e1,var="m1")
plot(boot.e1,var="m2")
```

10.3   Explore the Racial Disparity in Obesity

Lastly, we implement the method in a real data example: to explore the racial disparity in body mass index (BMI). The background of the data set is described in Section 7.4. We have found that on average, non-Hispanic blacks have a higher BMI and higher rate of obesity compared with non-Hispanic whites [85, 82] using the 2003-2006 National Health and Nutrition Examination Survey (NHANES). We are interested to see how the disparities can be explained by both individual and environmental factors. As a review, environmental risk factors are generated at the census tract level, which include both food environments and physical activity environments. Individual level variables include age, gender, smoking status, etc. Readers are referred to Section 7.4 for more details about the variables.

In this demonstration, we try to use risk factors from both levels to explain the racial disparity in BMI. We first use multiple additive regression trees to describe the relationships between BMI and all risk factors, and then we performed a data transformation on the risk factors so that the transformed variables have a roughly linear relationship with BMI. According to results from MART in Section 7.4, we did the following transformations for the individual level risk factors:

- The natural cubic spline bases for age with degrees of freedom of 4.

- Truncate the physical activity measurement to two parts: smaller than 2.1 and larger than 2.1 since we see a change point at 2.1 when depicting the relationship between physical activity and BMI.

We also use the natural cubic spline basis with different degrees of freedom on some of the environmental factors. The following codes shows how we make the transformations within the *mlma* package.

```
#dataset4 include the dataset
x<-dataset4[,"black"] # the vector of predictor
m4=dataset4[,c("Elevation","POP00_SQMI","mvper","male",
          "povcat","foreign","csmoker","hisp","age",
          "unhpopChi","Bars05_POP00","strden00",
"cnr01","intden00","Con_POP00_Chi")]
y4<-dataset4[,"bmxbmi"]    # outcome is bmi (continuous)

m4.2<-m4[,c("Elevation","PhysicalActivity","Male",
```

TABLE 10.1
Inferences on third-variable effects of risk factors in explaining racial disparities in BMI.

	Individual Level	Census Tract Level
Total Effect	$-1.94(-2.39,-1.07)$	$-0.09(-1.03,0.98)$
Race (direct effect)	$-1.46(-2.05,-0.72)$	$-0.08(-0.86,0.72)$
Age	$-0.30(-0.28,-0.17)$	$-0.01(-0.030,0.01)$
Foreign Born	$-0.11(-0.14,-0.07)$	$0.02(-0.01,0.04)$
Smoker	$0.06(0.03,0.09)$	$0.05(-0.02,0.11)$
Aex	$-.004(-.01,-.003)$	$-0.01(-.03,0.01)$
Physical Activity	$-0.08(-.01,-0.03)$	$-0.21(-0.39,-0.06)$
Elevation	$-$	$-0.04(-0.43,0.60)$
Street Density	$-$	$-0.07(-.28,0.18)$
Connected Node Ratio	$-$	$0.00(-0.00,0.00)$
Intersection Density	$-$	$0.25(-0.15,0.51)$

```
             "ForeignBorn","Smoker","age","StreetDensity",
             "ConnectedNodeCat","IntersectionDensity")]

data4.2<-data.org(x=x, levelx=1, m=m4.2, xref=1,
                l1=c("PhysicalActivity","age"),
                l2=c("Elevation", "StreetDensity",
                     "IntersectionDensity"),
                c1=c("Male","ForeignBorn","Smoker"),
                c1r=c(1,1,1), #set the reference group
                c2=c("ConnectedNodeCat"),
                c2r=c(0),
                f02ky=list(c("Elevation","StreetDensity",
                         "IntersectionDensity"),
                        c("ns(x,df=5)"),
                        c("ns(x,df=4)"),
                        c("ns(x,df=3)")),
                f20ky=list(c("age","PhysicalActivity"),
                        c("ns(x,df=4)"),
                        c("ifelse(x<2.1,x,2.1)",
                          "ifelse(x>2.1,x-2.1,0)")),
                level=fips)
```

The individual level exposure variable is the race (black (0) and white (1)). The census tract level exposure is the proportion of whites in the census tract. Table 10.1 shows the estimated third-variable effects at both the individual and census tract levels. Note that since high-level variable (e.g. census tract level) can influence the lower level (individual level) variable but not the reverse, all level one third-variables have both individual and census tract level effects, while all level two third-variables have only census tract level effects.

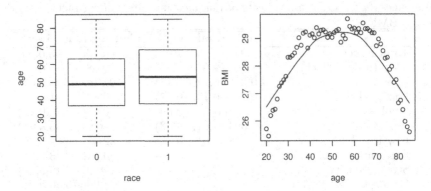

FIGURE 10.5
Indirect effect of age at the individual level.

We found that at the individual level, the average BMI for whites is 1.94 (TE) lower than that for blacks. Individual level factors can partially explain the racial difference. For example, age explains about 15% (−0.30 divided by −1.94) of the difference. Specifically, the left panel of Figure 10.5 shows that compared with whites, a larger proportion of blacks ($x = 0$) are at the middle age (late 30s to early 60s). In addition, the right panel shows that middle age group has the highest average BMI. The right panel of Figure 10.5 shows the relationship between age and BMI, which is not linear. BMI increases with age, peaked at the middle age and then declined. By the transformation of the third-variable "age" in analysis, we can catch the nonlinear relationship between age and BMI.

On the other hand, the racial difference in BMI at the census tract level is not significant. The total effect is −0.9 with a 95% confidence interval that includes 0. It is still important to use the multilevel models in the third-variable analysis to control for the correlations among people living in the same census tract.

11

Bayesian Third-Variable Effect Analysis

In this chapter, we discuss the use of the Bayesian method for the third-variable effect analysis. We provide example R codes that call WinBUGS (a software for Bayesian analysis) and perform the third-variable analysis for exposures, third-variables and outcomes of different formats.

11.1 Why Bayesian Method?

The Bayesian method is used widely in statistics inferences. The method is popular since it can use prior knowledge in addition to the currently collected data. Compared with the Frequentist method, parameters in Bayesian analysis are considered as random variables but not as fixed (unknown) numbers. Usually the purpose of the Bayesian analysis is to find out distributions of the parameters of interests given current data and prior knowledge. Under this background, denote θ as the vector of parameters of interest. With the Bayesian method, prior distributions are assigned to θ, say $P(\theta)$. Information from knowledge or previous data can be incorporated in the prior distributions. The information from current data are included in the likelihood function given the parameters, $L(\mathbf{X}|\theta)$. Given the data, the posterior distribution of parameters are deducted as $P(\theta|\mathbf{X}) \propto P(\theta) \times L(\mathbf{X}|\theta)$. The inferences for parameters are then based on the posterior distributions $P(\theta|\mathbf{X})$.

In general, two of the major benefits of using Bayesian inference are that: 1) The inferences are based on posterior distributions, therefore the interpretation of inferences are straightforward. For example, unlike the 95% confidence interval obtained from the Frequentist method, a 95% credible set built by the Bayesian method can be explained as that with 95% probability, the parameter is in the credible set. 2) Information from previous knowledge and data sets can be incorporated into the data analysis. Information gained from other research or knowledge is very important, especially when the sample size for the current study is small. When sample size becomes bigger, the information from current study dominates analysis results by the nature of the Bayesian method.

Specifically, the benefits of using Bayesian inference in the third-variable analysis are listed but not limited to the following aspects.

DOI: 10.1201/9780429346941-11

- Third-variables are random variables in the third-variable analysis, and they are being explained by exposure variable(s) and other covariates. This is naturally treated in the Bayesian models. Third-variables are assigned distributions whose parameters can depend on exposure(s) and covariates.

- Posterior distributions of third-variable effects are used to make inferences on the effects. There is no need of using bootstrap sampling to estimate the uncertainties of estimations.

- Due to the hierarchical structure of the Bayesian analysis, it is natural to take into account of the multilevel structure of data. Hence, it is straightforward to implement the multilevel third-variable analysis.

- It is easy to assign different error term structures in the Bayesian setting. Therefore, we can account for potential spatial and temporal correlations in the third-variable analysis.

- In the Frequentist analysis, separate models are fitted for the outcome and for each third-variable. In the Bayesian analysis, one final model is fitted for the outcome with hierarchical structures that take care of the dependence of third-variables on the exposure variable(s). Therefore, the Bayesian third-variable analysis can provide the measurement of goodness-of-fit (e.g. DIC) for the final model, which makes the model selection and comparison of the goodness-of-fits among different models straightforward.

There are drawbacks of using the Bayesian method in the third-variable analysis too. Mostly, the challenge might come from the computational expenses when there are a large number of third-variables. Current research has developed many computational methods for Bayesian analysis, such as Gibbs' sampler, Markov Chain Monte Carlo methods, etc. There are also a lot of Bayesian machine learning methods to help select variables and reduce dimension of important variables. Those methods can be adapted for the third-variable analysis.

In this chapter, we discuss the use of Bayesian method in the third-variable analysis. In the following sections, we first discuss the situation when the exposure variable is continuous. In Section 11.2, we introduce three different methods of estimating third-variable effects when both the outcome and third-variable(s) are continuous and then we extend these methods to deal with outcome and third-variable(s) of different formats (e.g. binary or discrete). In Sections 11.3, we deal with the third-variable effect analysis when the exposure variable is binary. The analysis when the exposure variable is multi-categorical or when there are multiple exposures is discussed in Section 11.4. All analysis methods in this chapter are implemented and presented in WinBUGS and R.

WinBUGS is a software that uses the Gibb's sampler to draw samples from posterior distributions of interests. For more details, readers are referred to [48]. To use WinBUGS, it has to be downloaded on the computer first. We show how to use WinBUGS for Bayesian third-variable analysis and how to interpret analysis results here. In this chapter, we focus on one third-variable only, but the codes provided can be easily extended for multiple third-variables.

11.2   Continuous Exposure Variable

In the Bayesian setting, we fit relationships among variables using generalized linear or additive models. Without loss of generality, we use non-informative prior distributions on the coefficients and error terms in the models. When knowledge is available, it can be incorporated into prior distributions through choices of different structure of prior distributions. Sensitivity analysis is sometimes required to show how the setting of informative priors can influence the final inferences.

We propose three methods for the third-variable effect analysis. Method one is based on the estimated coefficients. The third-variable effects are estimated directly as functions of observations and parameters from fitted models. The method has limited uses, but it generally useful for linear models. The second method is based on partial difference effect, where a direct effect is measured by the partial difference of the outcome on the exposure, and an indirect effect is measured by the product of the partial difference from the exposure on third-variables and that from third-variables to outcomes. The third method is based on a re-sampling scheme. The second and third methods are more general and can be used with any predictive model. We first illustrate the three methods when all variables are continuous and linear models are used to fit relationships among variables.

11.2.1   Continuous Outcome and Third-Variables

We start with a simple case where there is one third-variable, one exposure, and one outcome variable. All the variables are continuous. The following codes generate a simulation data set.

```
set.seed(1)
N=100
x=rnorm(N,0,1)
e1=rnorm(N,0,1)
e2=rnorm(N,0,1)
alpha=0.98
beta=1.96
c=1.28
M=alpha*x+e1
y=c*x+beta*M+e2
```

Example 11.1 *For the example, data are drawn from the following equations:*

$$M_i = 0.98X_i + \epsilon_{1i}$$
$$Y_i = 1.28X_i + 1.96M_i + \epsilon_{2i}$$

where X_i, ϵ_{1i} and ϵ_{2i} are standard normal distributed.

We use the R package *R2WinBUGS* to call WinBUGS from R and perform the MCMC sampling from the posterior distributions of parameters of interests. To use WinBUGS, first the WinBUGS package is downloaded from the website: https://www.mrc-bsu.cam.ac.uk/software/bugs/the-bugs-project-winbugs/ and unpacked in a local computer location, say "C:/folder1". The R package *R2WinBUGS* should also be downloaded and called in the R environment.

```
install.packages("R2WinBUGS")
library(R2WinBUGS)
```

11.2.1.1   Method 1: Functions of Estimated Coefficients

The first method estimates third-variable effects through the estimated coefficients from the predictive models. The method is useful when third-variable effects can be written as analytical functions of coefficients in the predictive models. In the above simulated data set, the indirect effect from M is *alpha* × *beta* and the direct effect is *c*. The following is the WinBUGS model for the analysis.

WinBUGS model for method 1 (noted as model_ccc1):

```
model {
#The model structure
for(i in 1:N){
mu_M[i] <- alpha*x[i]
M[i] ~ dnorm(mu_M[i],prec1)
mu_y[i] <-c*x[i]+beta*M[i]
y[i] ~ dnorm(mu_y[i],prec2)
}
#Set up prior distributions
alpha ~ dnorm(0.0,0.000001)
beta ~ dnorm(0.0,0.000001)
c ~ dnorm(0.0,0.000001)
var1 ~ dgamma(1,0.1)
var2 ~ dgamma(1,0.1)
prec1 <-1/var1
prec2 <-1/var2
#Calculation of third-variable effects
IE<-alpha*beta
```

```
DE<-c
}
```

In the model, N denotes the sample size, M is the vector of third-variable, y the vector of outcome and x the vector of the exposure variable. M, y and x are each of size $N \times 1$. Very vague priors are used for the coefficients and error terms. The indirect effect IE is defined to be *alpha* × *beta*, and the direct effect is c. The model is saved as a txt file with the file names "mediation_c_c_c_1.txt", in a local folder, say "c:/folder2". The following codes are executed in R to call WinBUGS for the third-variable analysis.

```
data1<-list(x=x,M=m,y=y,N=N)
inits<- function(){
 list(alpha=0,c=0,beta=0,prec1=0.5,prec2=0.5)
}
med1.1<-bugs(data1, inits, model.file =
            "C:/folder2/mediation_c_c_c_1.txt",
            parameters = c("alpha","c","beta",
         "var1","var2","IE","DE"),
            n.chains=1,n.iter=11000,n.burnin=1000,
            n.thin=1,bugs.directory =
 "C:/folder1/winbugs14_full_patched/WinBUGS14/",
            debug = F)
```

In the code, *data*1 defines the data to read into WinBUGS. *init* defines the initial values of parameters. In the *bugs* function, the argument *parameter* defines all parameters whose posterior samples are requested for monitoring. One chain is requested for the MCMC sampling where 11000 samples are drawn from the posterior distributions, 1000 of which are burn-in samples. The *debug* argument is set to be $TRUE$ if one would like to communicate with WinBUGS directly.

The results are stored in the *med*1.1 object. The results are printed as the following:

```
> med1.1
Inference for Bugs model at "C:/folder1/mediation_c_c_c_1.txt",
fit using WinBUGS, 1 chains, each with 11000 iterations (first
1000 discarded), n.sims = 10000 iterations saved
          mean   sd  2.5%   25%   50%   75% 97.5%
alpha      1.0  0.1   0.8   0.9   1.0   1.0   1.2
c          1.3  0.2   1.0   1.2   1.4   1.5   1.6
beta       1.9  0.1   1.7   1.8   1.9   2.0   2.1
var1       1.0  0.1   0.7   0.9   0.9   1.0   1.3
var2       1.1  0.2   0.9   1.0   1.1   1.2   1.5
IE         1.9  0.2   1.4   1.7   1.9   2.0   2.3
DE         1.3  0.2   1.0   1.2   1.4   1.5   1.6
deviance 568.8  3.3 564.4 566.4 568.2 570.5 577.1
```

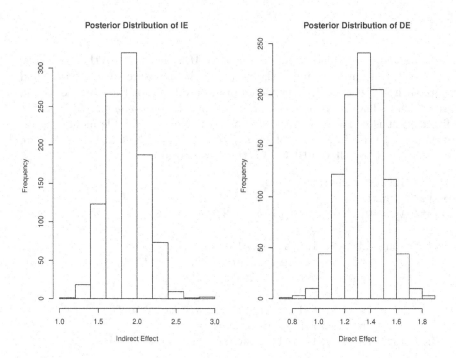

FIGURE 11.1
The posterior distributions of the direct and indirect effects.

```
DIC info (using the rule, pD = Dbar-Dhat)
pD = 4.9 and DIC = 573.7
DIC is an estimate of expected predictive error (lower
deviance is better).
```

We found that the estimated indirect effect from M is 1.9 ($sd = 0.2$) and the estimated direct effect is 1.3 ($sd = 0.2$). Both are very close to the true values. The posterior samples are stored as a matrix in $med1.1\$sims.matrix$. Researchers can use the samples directly. Figure 11.1 shows the posterior distributions of the direct (DE) and indirect effects (IE) out of the samples. Note that the posterior distribution of the direct effect is almost normal, but the posterior distribution of the indirect effect is not symmetric around the mean and therefore does not have an exact normal distribution. The indirect effect is the product of two random variables each of which has a normal distribution.

11.2.1.2   Method 2: Product of Partial Differences

The second method is to calculate third-variable effects through partial differences. If the estimation function of the outcome, $E(Y) = f(X, m)$ is differentiable based on the exposure variable X and the third-variable(s) M. The direct effect of the exposure variable is defined as the partial changing rate in the outcome when the exposure variable changes. Therefore it can be calculated by that $DE(x) = \frac{\partial f(X,M)}{\partial X}|_{X=x}$.

If the estimation function of the exposure variable on the third-variable M, $E(M) = g(x)$, is also differentiable, by definition, the indirect effect from X to Y through M is $IE_M(x) = \frac{\partial g(X)}{\partial X} \cdot \frac{\partial f(X,M)}{\partial M}|_{X=x}$. The interpretation of the indirect effect is straightforward: one unit change in X result in $\frac{\partial g(X)}{\partial X}$ units of change in the expected M, and one unit of change in M results in $\frac{\partial f(X,M)}{\partial M}$ units of changes in the expected outcome. Therefore the changing rate of the outcome through M is the product of the two partial differentiations.

To make the method more general without the needs to do differentiation for different models, we replace the partial differentiations with the partial differences. Therefore, $DE(x)$ is calculated as $\frac{f(x+\Delta x,M)-f(x,M)}{\Delta x}$ and $IE(x)$ as $\frac{g(x+\Delta x)-g(x)}{\Delta x} \cdot \frac{f(x,M+\Delta m)-f(x,M)}{\Delta m}$.

Finally for the choices of Δx and Δm, too big values make the approximation of the differences to the differentiations to be apart. We suggest to choose Δx and Δm as the minimum of 0.01 or the range of X or M divided by 100.

WinBUGS model for method 2 (noted as model_ccc2):

```
model {
for(i in 1:N){
mu_M[i] <- alpha*x[i]
M[i] ~ dnorm(mu_M[i],prec1)
mu_y[i]<- c*x[i]+beta*M[i]
y[i] ~ dnorm(mu_y[i],prec2)

mu_y1[i]<- c*(x[i]+deltax)+beta*M[i]
de[i]<-(mu_y1[i]-mu_y[i])/deltax

mu_M1[i] <- alpha*(x[i]+deltax)
mu_y2[i]<- c*x[i]+beta*(M[i]+deltam)
ie[i]<-(mu_M1[i]-mu_M[i])/deltax*(mu_y2[i]-mu_y[i])/deltam

te[i]<-ie[i]+de[i]
}

alpha ~ dnorm(0.0,0.000001)
beta ~ dnorm(0.0,0.000001)
c ~ dnorm(0.0,0.000001)
var1 ~ dgamma(1,0.1)
var2 ~ dgamma(1,0.1)
prec1 <-1/var1
prec2 <-1/var2
}
```

To check how the code works with nonlinear relationships, we generate the following third-variable $M.2$ and outcome $y.2$. All other variables are like the original simulation.

Example 11.2 *For the example, the data are drawn from the following equations:*

$$M.2_i = 0.98X_i^2 + \epsilon_{1i}$$
$$Y.2_i = 1.28X_i + 1.96M.2_i + \epsilon_{2i}$$

where X_i, ϵ_{1i} and ϵ_{2i} are standard normal distributed.

To run the Bayesian third-variable analysis, we change the lines 3 and 9 in the model model_ccc2 according to the changed model $g(x)$. Line 3 is changed to $mu_M[i] < -alpha * x[i] * x[i]$ and line 9 should be $mu_M1[i] < -alpha * (x[i] + deltax) * (x[i] + deltax)$. All other codes are the same. Run the R codes and the estimated average indirect effect becomes 0.39, which is very close to $2alpha \cdot beta \cdot mean(X)$. The estimated indirect effect by X is shown in the left panel of Figure 11.2, which is not constant to X, but is $2alpha \cdot beta \cdot x$, due to the quadratic relationship between X and M.

11.2.1.3   Method 3: A Resampling Method

The third Bayesian method we propose here is similar to the Frequentist method presented in previous chapters. The total effect is the changing rate in the outcome when the exposure variable changes, which results in changes in all third-variables. Using notations from above, let $M(x)$ be the random variable of M given X. The total effect, $TE(x)$ is calculated as $\frac{f(x+\Delta x, M(x+\Delta x)) - f(x, M(x))}{\Delta x}$.

The direct effect not from a certain third-variable, M, is that when the exposure variable changes, all other third-variables change with X except for M. The resulted changing rate in the outcome is called the direct effect not from M. And the difference between the total effect and the direct effect not from M is the indirect effect from M. When allowing M changes with the exposure variable, we draw $M(x)$ from the conditional distribution of M given $X = x$. When M does not change with X, each time we randomly draw an X and then draw M conditional on the random X. Therefore, M is drawn from its marginal distribution. The direct effect not from M is calculated as $\frac{f(x+\Delta x, M) - f(x, M)}{\Delta x}$. Besides M, if there are other third-variables, denoted as $\tilde{M}$, the direct effect not from M is calculated as $\frac{f(x+\Delta x, M, \tilde{M}(x+\Delta x)) - f(x, M, \tilde{M}(x))}{\Delta x}$.

For the choice of Δx, we want it to be small for a good approximation of the differences to the differentials. However, if Δx is too small, the estimation results can be very variant since Δx is used as the denominator in the estimation and there are randomness in the numerator brought in by sampling from marginal distributions. In the example, we choose Δx as that for method 2: $\min(0.01, range(X)/100)$. The iteration times of MCMC should be

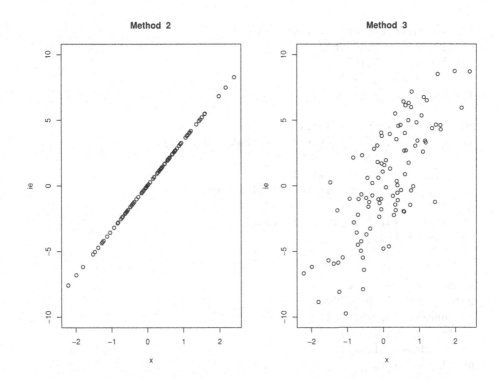

FIGURE 11.2
The posterior mean of the indirect effect of M by the exposure variable (X) by method 2 (left) and method 3 (right).

large enough to ensure the convergence of results. The following is the Win-BUGS model for the third method on Example 11.1, when there is only one third-variable that is linearly related with X and with Y.
WinBUGS model for method 3 (model_ccc3):

```
model {
for(i in 1:N){
mu_M[i] <- alpha*x[i]
M[i] ~ dnorm(mu_M[i],prec1)

mu_y[i]<- c*x[i]+beta*M[i]
y[i] ~ dnorm(mu_y[i],prec2)

mu_y0[i]<-c*x[i]+beta*mu_M[i]
mu_M1[i] <- alpha*(x[i]+deltax)
mu_y1[i]<- c*(x[i]+deltax)+beta*mu_M1[i]
te[i]<-(mu_y1[i]-mu_y0[i])/deltax
```

```
#draw random numbers between 1 and N
j1[i]~dunif(0.5,100.5)
j2[i]~dunif(0.5,100.5)
j3[i]<- round(j1[i])
j4[i]<- round(j2[i])

mu_M2[i] <- alpha*(x[j3[i]])
mu_y2[i]<- mu_y0[i]-beta*mu_M[i]+beta*mu_M2[i]

mu_M3[i] <- alpha*(x[j4[i]])
mu_y3[i]<- mu_y1[i]-beta*mu_M1[i]+beta*mu_M3[i]

de[i]<-(mu_y3[i]-mu_y2[i])/deltax
ie[i]<-te[i]-de[i]
}

alpha ~ dnorm(0.0,0.000001)
beta ~ dnorm(0.0,0.000001)
c ~ dnorm(0.0,0.000001)
var1 ~ dgamma(1,0.1)
var2 ~ dgamma(1,0.1)
prec1 <-1/var1
prec2 <-1/var2
}
```

The model can be modified for nonlinear relationships. For Example 11.2, we only need to change the model for M and apply it on $M.2$ and $y.2$. The estimated indirect effect of $M.2$ by the exposure variable is drawn on the right panel of Figure 11.2. The estimates from method 3 is more variant than that from method 2. This is due to the randomness brought in by drawing from the marginal distribution of $M.2$.

11.2.2   Different Format of Outcome and Third-Variables

In this section, we extend the all-continuous-variables situation to handle different types of outcomes (Section 11.2.2.1) and/or third-variables (Sections 11.2.2.2 and 11.2.2.3).

11.2.2.1   Outcomes of Different Format

In general, Bayesian models can be easily extended to generalized linear and additive models. We basically change the original linear model by using a different random term and a link function. For example, if the outcome is binary and we decide to use a logistic regression model to fit the outcome, we change the model for the outcome from

```
mu_y[i]<- c*x[i]+beta*M[i]
y[i] ~ dnorm(mu_y[i],prec2)
```

to

```
logit(mu_y[i]) <- c*x[i]+beta*M[i]
y[i] ~ dbern(mu_y[i])
```

Another example is when the outcome is discrete and we use a log-linear model to fit it. The model for the outcome is of the form:

```
log(mu_y[i]) <-  c*x[i]+beta*M[i]+b
y[i] ~ dpois(mu_y[i] )
```

The three methods described for continuous outcomes in Section 11.2.1 can be used for outcomes of different formats. Using the binary outcome as an example. Assume that a logistic regression is used to fit the relationship between the outcome and other variables.

Example 11.3 *For the example, all other variables are drawn as in Example 11.1 except for the outcome is binary from the model*

$$logit(Y_i) = 1.28X_i + 1.96M_i.$$

If method 1 is used, the indirect effect can still be *alpha* × *beta*, however, the interpretation of the indirect effect is different. The third-variable effects are in terms of the log-odds, but not directly on the mean (the probability) of the outcome. For example, the direct effect, DE, is explained as that when X increases by one unit, the odds of $Y = 1$ becomes $exp(DE)$ times of the original odds directly through X. Compared with method 1, methods 2 and 3 can calculate the third-variable effects in terms of the changing rate in the mean of the outcome when the exposure variable changes. Therefore, the interpretation of the third-variable effect is still in terms of the mean of the outcome, i.e. $P(Y = 1)$. Since the link function is not linear, the direct effect and the indirect effect are not constant over different values of the exposure variable. If the multiplicative effect is preferred, the change rate should be calculated in terms of the logit mean of the outcome, but not of the original format of the mean. In the book website, we provide the R codes and WinBUGs models for binary outcomes.

11.2.2.2 Binary Third-Variables

When a third-variable is of different format, generalized linear models can be used to fit the relationship between the third-variable and the exposure variable. As for the outcome, we need to change the random term and give a link function that links the linear components with the mean of the third-variable. For example, if M is binary and we fit a logistic regression for the relationship between M and X, the model in WinBUGS is listed as the following:

```
logit(mu_M[i]) <- alpha*x[i]
M[i] ~ dbern(mu_M[i])
```

For such case, the *alpha* × *beta* in method 1 cannot be directly used to calculate the indirect effect. Since in the model, a unit change in X is related with a multiplicative change in the odds of $M = 1$, *exp(alpha)* times the original odds, but not with an additive change in the mean of the third-variable. Therefore, although c is still the direct effect of X, *alpha* × *beta* is not the indirect effect.

Methods 2 and 3 are based on the changing rate in the mean of the third-variable. Therefore both methods can be used directly for different formats of the third-variable. To illustrate the method, we simulate a dataset with a binary third-variable. Other variables are generated as the first example except for the outcome and the third-variable are generated using the following codes:

```
mu_m=alpha*x
M3=rbinom(100,1,exp(mu_m)/(1+exp(mu_m)))
y2.1=c*x+beta*M3+e2
```

Example 11.4 *For the example, all other variables are drawn as in Example 11.1 except for the third-variable is binary from the model*

$$logit(M_i) = 0.98X_i.$$

Using the second method, the following WinBUGs model is stored in the file "mediation_c_b_c_2.txt" in the folder "C:/folder2". *WinBUGS model for method 2 with binary third-variable (model_cbc2):*

```
model {
for(i in 1:N){
logit(mu_M[i]) <- alpha*x[i]
M[i] ~ dbern(mu_M[i])
mu_y[i]<- c*x[i]+beta*M[i]
y[i] ~ dnorm(mu_y[i],prec2)

logit(mu_M1[i]) <- alpha*(x[i]+deltax)
M1[i] ~ dbern(mu_M1[i])
mu_y1[i]<- c*(x[i]+deltax)+beta*M[i]
mu_y2[i]<- c*x[i]
mu_y3[i]<- c*x[i]+beta

ie[i]<-(mu_M1[i]-mu_M[i])/deltax*(mu_y3[i]-mu_y2[i])/deltam
de[i]<-(mu_y1[i]-mu_y[i])/deltax
te[i]<-ie[i]+de[i]
}

alpha ~ dnorm(0.0,0.01)
beta ~ dnorm(0.0,0.01)
c ~ dnorm(0.0,0.001)
```

```
var2 ~ dgamma(1,0.1)
prec2 <-1/var2
}
```

The following code is called in R to run the model:

```
deltax=diff(range(x))/10
deltam=1
data1<-list(x=x,M=M3,y=y2.1,N=N,deltax=deltax,deltam=deltam)
inits<-function(){list(alpha=0,c=0,beta=0,prec2=0.5)}
med3.2<- bugs(data1, inits, parameters = c("alpha","c","beta",
              "var2","ie","de","te"),
              model.file = "C:/folder2/mediation_c_b_c_2.txt",
 n.chains = 1, n.iter = 11000,n.burnin=1000,
              bugs.directory = "C:/folder1/WinBUGS14/",
              debug = F)
```

Left panel of Figure 11.3 shows the indirect effect of M versus X using method 2. Because of the logit link between the mean of M and $alpha \times x$. The indirect effect of M is not constant over X. The third method can be similarly implemented. The estimated indirect effects versus X using method 3 is shown in the right panel of Figure 11.3. We provide the codes in the book website.

11.2.2.3 Categorical Third-Variables

When a third-variable is categorical, the indirect effect from it needs to include the changes in the probabilities of all levels when the exposure changes. In the following we use an example to illustrate how to estimate the indirect effect from a third-variable of three levels. The codes below generate a third-variable of three levels based on the continuous exposure variable that is generated in Example 11.1. In the code, $M4$ is the third-variable with three levels. It is of dimension $N \times 3$, where each row is an observation and each column indicates the category of the observation. For example, if the ith observation is of the second category, the ith row of $M4$ is $(0, 1, 0)$. $M4.1$ is a vector that indicates the category of observations, e.g. $M4.1[i] = 2$. $M4.1$ is going to be read into WinBUGS for analysis.

```
set.seed(2)
alpha1=0.5
alpha2=0.8
beta1=1
beta2=1.2
mu_m1=alpha1*x
mu_m2=alpha2*x
p=cbind(1,exp(mu_m1),exp(mu_m2))
M4=t(apply(p,1,rmultinom,n=1,size=1))
y4=c*x+beta1*M4[,2]+beta2*M4[,3]+e2
M4.1=as.vector(M4%*%(1:3))
```

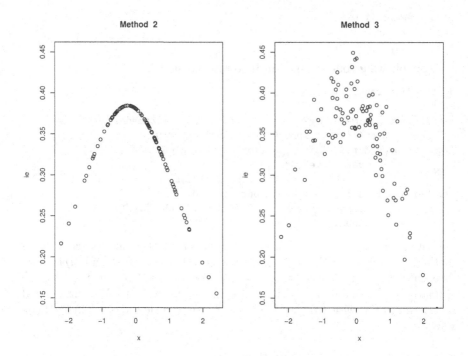

FIGURE 11.3
The estimated indirect effects of M versus the exposure variable using method 2 (left) and method 3 (right).

Example 11.5 *For the example, all other variables are drawn as in Example 11.1 except for the third-variable, $M4$, has three levels, where the probability of each level depends on the exposure variable x in that:*

$$\log \frac{P(m4_i = 2)}{P(M4_i = 1)} = 0.5x_i$$

$$\log \frac{P(m4_i = 3)}{P(M4_i = 1)} = 0.8x_i$$

The outcome variable is generated by:

$$y_i = 1.28x_i + I(m4_i = 2) + 1.2I(m4_i = 3) + \epsilon_2$$

$I(\cdot)$ is an indicator function that equals 1 if the $\cdot$ is true and 0 otherwise. The first level of $m4_i$ is used as the reference level.

In this situation, method 1 is still useful if we can write the indirect effect of $M4$ directly as a function of the parameters. The code and WinBUGS model is given in the book website. But method one is not general since for different cases, the function for indirect effect can be very different. Method 2 and 3 can be easily generalized to other cases. The WinBUGS model, "mediation_c_cat_c_2" for method 2 is listed as follows:

```
model {
for(i in 1:N){
mu_M1[i,1] <- 1  #baseline is the 1st category
mu_M1[i,2] <- exp(alpha1*x[i])
mu_M1[i,3] <- exp(alpha2*x[i])
sum_M[i] <-mu_M1[i,1]+mu_M1[i,2]+mu_M1[i,3]
for (k in 1:3)
{mu_M[i,k] <- mu_M1[i,k]/sum_M[i]}
M[i]~dcat(mu_M[i,])
mu_y[i] <-c*x[i]+beta1*equals(M[i],2)+beta2*equals(M[i],3)
y[i] ~ dnorm(mu_y[i],prec2)

mu_y1[i]<- c*(x[i]+deltax)+beta1*equals(M[i],2)+beta2*equals(M[i],3)
de[i]<-(mu_y1[i]-mu_y[i])/deltax

mu_M1.2[i,1] <- 1  #baseline is the 1st category
mu_M1.2[i,2] <- exp(alpha1*(x[i]+deltax))
mu_M1.2[i,3] <- exp(alpha2*(x[i]+deltax))
sum_M.2[i] <-mu_M1.2[i,1]+mu_M1.2[i,2]+mu_M1.2[i,3]
for (k in 1:3)
{mu_M.2[i,k] <- mu_M1.2[i,k]/sum_M.2[i]}

mu_y2[i]<- c*x[i]
mu_y3[i]<- c*x[i]+beta1
mu_y4[i]<- c*x[i]+beta2

ie1[i]<-(mu_M.2[i,2]-mu_M[i,2])/deltax*(mu_y3[i]-mu_y2[i])/deltam1
ie2[i]<-(mu_M.2[i,3]-mu_M[i,3])/deltax*(mu_y4[i]-mu_y2[i])/deltam2
ie[i]<-ie1[i]+ie2[i]
te[i]<-ie[i]+de[i]
}

alpha1 ~ dnorm(0.0,0.01)
alpha2 ~ dnorm(0.0,0.01)
beta1 ~ dnorm(0.0,0.000001)
beta2 ~ dnorm(0.0,0.000001)
c ~ dnorm(0.0,0.001)
var2 ~ dgamma(1,0.1)
prec2 <-1/var2
}
```

In the model, we separate the indirect effect by the different level of $M4$. The $ie1$ in model is the changing rate in Y through the change in the probability of $M4 = 2$ when X changes. $ie2$ is the changing through the change in the $P(M4 = 3)$. The indirect effect is the summation of the separated

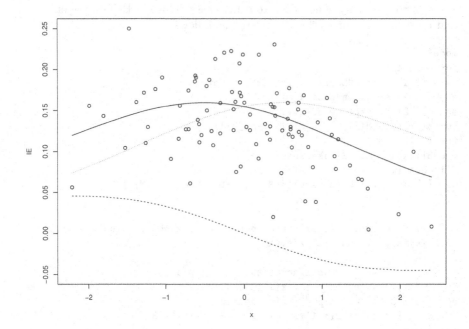

FIGURE 11.4
The estimated indirect effects of the categorical third-variable $M4$ through
method 2 (the solid line) and method 3 (the dots).

indirect effects. In the same way, we can use the method to find group effects
from multiple third variables. In Figure 11.4, the solid line is the estimated
indirect effect from $M4$, which is decomposed into effects from the change of
$P(M4 = 2)$ (the dashed line) and the change of $P(M4 = 3)$ (the dotted line).

Method 3 is also used for Example 11.5. Compared with method 2, the
WinBUGS modle, denoted as "mediation_c_cat_c_3", is as follows:

```
for(i in 1:N){
mu_M1[i,1] <- 1  #baseline is the 1st category
mu_M1[i,2] <- exp(alpha1*x[i])
mu_M1[i,3] <- exp(alpha2*x[i])
sum_M[i] <-mu_M1[i,1]+mu_M1[i,2]+mu_M1[i,3]
for (k in 1:3)
{mu_M[i,k] <- mu_M1[i,k]/sum_M[i]}
M[i]~dcat(mu_M[i,])
mu_y[i] <-c*x[i]+beta1*equals(M[i],2)+beta2*equals(M[i],3)
y[i] ~ dnorm(mu_y[i],prec2)
```

```
j1[i]~dunif(0.5,100.5)
j2[i]~dunif(0.5,100.5)
j3[i]<- round(j1[i])
j4[i]<- round(j2[i])

mu_M1.2[i,1] <- 1  #baseline is the 1st category
mu_M1.2[i,2] <- exp(alpha1*(x[i]+deltax))
mu_M1.2[i,3] <- exp(alpha2*(x[i]+deltax))
sum_M.2[i] <-mu_M1.2[i,1]+mu_M1.2[i,2]+mu_M1.2[i,3]
for (k in 1:3)
{mu_M.2[i,k] <- mu_M1.2[i,k]/sum_M.2[i]}
mu_y1[i]<- c*(x[i]+deltax)+beta1*mu_M.2[i,2]+beta2*mu_M.2[i,3]
mu_y0[i] <-c*x[i]+beta1*mu_M[i,2]+beta2*mu_M[i,3]
te[i]<-(mu_y1[i]-mu_y0[i])/deltax

mu_M1.3[i,1] <- 1  #baseline is the 1st category
mu_M1.3[i,2] <- exp(alpha1*x[j3[i]])
mu_M1.3[i,3] <- exp(alpha2*x[j3[i]])
sum_M.3[i] <- mu_M1.3[i,1]+mu_M1.3[i,2]+mu_M1.3[i,3]
for (k in 1:3)
{mu_M.3[i,k] <- mu_M1.3[i,k]/sum_M.3[i]}
mu_y2[i]<- c*x[i]+beta1*mu_M.3[i,2]+beta2*mu_M.3[i,3]

mu_M1.4[i,1] <- 1  #baseline is the 1st category
#changed j4 to j3 to reduce variance
mu_M1.4[i,2] <- exp(alpha1*x[j4[i]])
mu_M1.4[i,3] <- exp(alpha2*x[j4[i]])
sum_M.4[i] <- mu_M1.4[i,1]+mu_M1.4[i,2]+mu_M1.4[i,3]
for (k in 1:3)
{mu_M.4[i,k] <- mu_M1.4[i,k]/sum_M.4[i]}
mu_y3[i]<- c*(x[i]+deltax)+beta1*mu_M.4[i,2]+beta2*mu_M.4[i,3]
de[i]<-(mu_y3[i]-mu_y2[i])/deltax
ie[i]<-te[i]-de[i]
}
```

The dots in Figure 11.4 are the results of estimated indirect effect of $M4$ through Method 3. Again, the estimated indirect effect is a little more variant, which is brought in by the randomness of sampling from the marginal distribution of the third variable. To reduce the variance, we can change the $j4$ in the model to $j3$.

11.3 Binary Exposure Variable

In this section, we discuss the Bayesian third-variable analysis method with binary exposures. Since the exposure variable can take only two potential values, taking differential on the exposure variable is not possible. However, it is still reasonable to take differences with respect to the two conditional expectations of the third-variable. We discuss the binary exposure variable with continuous outcome and third-variables in this section. As discussed in Section 11.2.2, the methods used for continuous outcome and third-variables can be easily extended for other variable formats.

Without loss of generosity, assume that the binary outcome is transformed and takes the value of 0 or 1, where $X = 0$ is used as the reference group. If method 1 is used for the analysis, the third-variable effects can be calculated using the model ccc_1. No change is needed in the model for the binary exposure.

For method 2, the model changes a little where deltax is set to be 1 and X is 0 when calculating the difference in X. Specifically, model_ccb2 is used for Example 11.1 with method 2 and a binary exposure.

WinBUGS model for method 2 with binary exposure (model_ccb2):

```
model {
for(i in 1:N){

mu_M[i] <- alpha*x[i]
M[i] ~ dnorm(mu_M[i],prec1)
mu_y[i]<- c*x[i]+beta*M[i]
y[i] ~ dnorm(mu_y[i],prec2)

mu_y3[i]<-beta*M[i] #when x=0
mu_y4[i]<-c+beta*M[i] #when x=1
de[i]<-(mu_y4[i]-mu_y3[i])/deltax

mu_y2[i]<-c*x[i]+beta*(M[i]+deltam)
ie[i]<-alpha/deltax*(mu_y2[i]-mu_y[i])/deltam #alpha is alpha(1-0)

te[i]<-ie[i]+de[i]
}

alpha ~ dnorm(0.0,0.000001)
beta ~ dnorm(0.0,0.000001)
c ~ dnorm(0.0,0.000001)
var1 ~ dgamma(1,0.1)
var2 ~ dgamma(1,0.1)
prec1 <-1/var1
prec2 <-1/var2
}
```

The change in the model for method 3 is similar:

WinBUGS model for method 3 with binary exposure (model_ccb3):

```
model {
for(i in 1:N){
mu_M[i] <- alpha*x[i]
```

```
M[i] ~ dnorm(mu_M[i],prec1)

mu_y[i]<- c*x[i]+beta*M[i]
y[i] ~ dnorm(mu_y[i],prec2)

j1[i]~dunif(0.5,100.5)
j2[i]~dunif(0.5,100.5)
j3[i]<- round(j1[i])
j4[i]<- round(j2[i])

M1[i] ~ dnorm(alpha,prec1)

M0[i] ~ dnorm(0,prec1)

mu_y1[i]<- c+beta*M1[i]
mu_y0[i]<- beta*M0[i]
te[i]<-(mu_y1[i]-mu_y0[i])/deltax

mu_M2[i] <- alpha*(x[j3[i]])
M2[i] ~ dnorm(mu_M2[i],prec1)
mu_y2[i]<- beta*M2[i]

mu_M3[i] <- alpha*(x[j4[i]])
M3[i] ~ dnorm(mu_M3[i],prec1)
mu_y3[i]<- c+beta*M3[i]

de[i]<-(mu_y3[i]-mu_y2[i])/deltax
ie[i]<-te[i]-de[i]
}

alpha ~ dnorm(0.0,0.000001)
beta ~ dnorm(0.0,0.000001)
c ~ dnorm(0.0,0.000001)
var1 ~ dgamma(1,0.1)
var2 ~ dgamma(1,0.1)
prec1 <-1/var1
prec2 <-1/var2
}
```

The codes and a simulation study are provided in the book website.

11.4 Multiple Exposure Variables and Multivariate Outcomes

When there are multiple exposure variables and/or multivariate outcomes, we need to calculate a set of third-variable effects for each pair of the exposure-outcome relationship. That is, if we have E exposure variables and O outcomes, we need to estimate the sets of third-variable effects for the $E \times O$ pairs of exposure-outcome relationships. When an exposure variable or an outcome is multi-categorical of K categories, we handle that by transforming the variable into $K - 1$ binary variables. Then the transformed variables are treated as multiple exposures or multivariate outcomes in the third-variable analysis. We need to keep the correlations among the multivariate outcomes in the third-variable analysis. This is easily handled in the Bayesian setting by assigning specific prior distributions and likelihood functions among variables.

12

Other Issues

In this chapter, we discuss important issues including explaining third-variable effects (Section 12.1), the power analysis and sample size calculation (Section 12.2), and the strategies for sequential and longitudinal third-variable analysis (Section 12.3).

12.1   Explaining Third-Variable Effects

Causal Effect or Not. As discussed in previous chapters, if there are causal relationships from the exposure to a third-variable and from the third-variable to the outcome, the effect from the third-variable is interpreted as a mediation effect. Otherwise, the effect may be explained as a confounding effect.

According to the purpose of analysis, the third-variable effect could be mediation or confounding effect. In the field of health disparities, third-variable effect analysis tends to be confounding analysis. For example, if the purpose is to explore risk factors that can explain the racial difference in breast cancer survival (see Sections 5.3.2 and 5.3.3), since race does not directly change risk factors, the analysis is to find confounding effects that can explain the racial difference. Therefore effects from third-variables are mostly confounding effect. For example, African Americans have a lower private insurance coverage, which is related with worse survival rate. If a confounding effect is found significantly explain the health disparity, interventions can be made to change the relationship. In the above example, the intervention could aim at increasing the coverage of health insurance among African Americans.

On the other side, the third-variable analyses in clinical trials are more likely to be mediation analysis. For example, if the purpose of analysis is to find out how a treatment works to reduce obesity. It is hypothesized that the treatment works by reducing food-intake or by increasing physical activities. By directly changing individual behaviors, the treatment has indirect effects on reducing the chance of being obese. Therefore, the third-variable analysis involving the treatment as an exposure variable on how it works to reduce weight is a mediation analysis.

In an exposure-outcome relationship, it is possible that third-variables involve both mediators and confounders. Depending on the analysis purpose,

DOI: 10.1201/9780429346941-12   245 the confounders enter the third-variable analysis either as covariates to the outcome only, or as third-variables. If the purpose is to explore what causes the exposure-outcome relationship, confounders should be included as covariates. If the research aims at explaining the associations between the exposure and outcome, confounders can be included as third-variables. Theoretically, the indirect effect from a specific mediator would not change between the two methods. The only difference in results would be whether the confounding effects are included in the total effect nor not.

Negative Relative Effects. The relative effect is defined as a direct or indirect effect divided by the total effect. The total effect is all effects of paths between the exposure and the outcome. A relative effect could be negative if the (in)direct effect has a different sign compared to the total effect. For example, the total effect is positive and the (in)direct effect is negative. In such situation, the third-variable is sometimes called a depressive factor, meaning that instead of explaining the exposure-outcome relationship, the third-variable actually withholds it. That is, controlling the third-variable could enhance rather than explain the association between the exposure and the outcome. If the relative direct effect is negative, the third-variables included in the analysis "over-explain" the exposure-outcome relationship.

A total effect closing to 0 could be the result of the existence of both negative and positive effects on different paths. In such case, the absolute values of relative effects could be very large, e.g. larger than 100%. Therefore, the relative effect is more reasonable and interpretable if mostly the direct effect and indirect effects are of the same direction.

Group Effects. Using methods proposed in the book, effects of groups of third-variables can be calculated. It is not necessary that the group effect equals the sum of effects from all third-variables even when linear models are fitted. This is because potential col-linearities among the third-variables. Specially, the upper or lower bound of the group effect is not likely to be the sum of the bounds for each third-variable in the group. It is possible that all third-variables are not significant while the group effect is significant or the reverse.

This is similar for multi-categorical third-variables. As in the third-variable analysis, a multi-categorical third-variable of K levels is generally decomposed into $K-1$ binary variables. The indirect effect through each binarized variable is calculated. The indirect effect of the original third-variable is the group effect of the binarized variables. Since the binarized variables are orthogonal to each other, collinearity is not likely to happen. However, it can still happen that each binarized variable does not have a significant indirect effect, while the original multi-categorical variable has.

Average Effects. The average effect refers to the average direct, indirect or total effect over the domain of the exposure variable/predictor. When there are nonlinear relationships among variables and when the exposure variable is continuous, it is possible that the third-variable effect at a certain value of the exposure variable is different from the average third-variable effect.

There are many such examples in the book. It is therefore possible that the third-variable effect is significant at a specific value or range of values of the exposure variable, while the average third-variable effect is not significant. For example, the third-variable effect could be negative in a range and positive in another range, thereby the average third-variable effect is not significantly different from 0. When this happens, it is incorrect to conclude that the third-variable effect is insignificant. The third-variable effect over the whole domain of the exposure variable is informative and should be reported. All R packages created for the third-variable analysis in the book provide tools to report and depict third-variable effects over the domain of every exposure variable.

12.2 Power Analysis and Sample Sizes

In the power analysis for third-variable effect inferences, we would like to find out the sample size needed to identify a significant indirect effect from a third-variable or to identify a significant direct effect after adjusting for other variables. The latter, identifying a significant direct effect, is straightforward since it involves only one model that we would like to check the significance of the variable (the exposure) when other variables, third-variables and co-variates, are adjusted. Depending on the format of the final model, the power analysis can be performed using established method or software (e.g. PASS). However, the former power analysis, to identify significant third-variables, involves at least two models and in the simplest situation, the product of two estimated effects from different models. This makes the power analysis more complicated. In the following subsections, we discuss two methods for the power analysis of identifying significant third-variables. One method is based on finding the distribution of the product of two random variables and the other is based on Monte Carlo method.

12.2.1 Linear Models

Using linear models, the third-variable analysis involves at least two models to identify an indirect effect. Take the example that both the third-variable and the outcome are continuous and their relationships with the exposure variable and other covariates are fitted using linear regressions. The estimated indirect effect is represented in the form of the product of two normal distributed random variables. To perform the power analysis of identifying a significant indirect effect, we need to know the distribution of the indirect effect estimand. In the linear regression setting, denote $\hat{a}$ as the fitted coefficient for the exposure variable, X, when it is used to estimate the third-variable M and $\hat{b}$ as the coefficient for the third-variable, M, when it is used to estimate the outcome Y adjusting for the exposure(s) and other variables. The estimated indirect effect of M is thus $\hat{a} \times \hat{b}$. By the assumptions of linear regressions, both $\hat{a}$ and $\hat{b}$ have normal distributions. We would like to find out the distribution of $\hat{a} \times \hat{b}$ to calculate the power of finding that $\hat{a} \times \hat{b}$ is significantly different from zero.

Assume that $\hat{a}$ has a normal distribution with mean a and variance σ_a^2 and $\hat{b}$ has a normal distribution with mean b and variance σ_b^2. First, we assume $\hat{a}$ and $\hat{b}$ are independent. This might be true if given M, X and Y are independent. That is, M completely explain the $X - Y$ association. We denote $Z = \hat{a} \times \hat{b}$. The moment generating function of Z is then

$$M_Z(t) = \frac{exp\{\frac{tab+\frac{1}{2}(b^2\sigma_a^2+a^2\sigma_b^2)t^2}{1-t^2\sigma_a^2\sigma_b^2}\}}{\sqrt{1-t^2\sigma_a^2\sigma_b^2}}.$$

Now we define a new parameter $\delta_{ab} = \sigma_a\sigma_b$, and replace it in the above equation, the moment generating function becomes:

$$M_Z(t) = \frac{exp\{\frac{tab+\frac{1}{2}(b^2\sigma_a^2+a^2\sigma_b^2)t^2}{1-t^2\delta_{ab}^2}\}}{\sqrt{1-t^2\delta_{ab}^2}}.$$

When δ_{ab} goes to 0, which happens when the sample size increases and at least one of the variances for $\hat{a}$ or $\hat{b}$ goes to 0, the limiting moment generating function goes to

$$\lim_{\delta_{ab}\to 0} M_Z(t) = exp\{tab + \frac{1}{2}(b^2\sigma_a^2 + a^2\sigma_b^2)t^2\},$$

which is the moment generating function of a random variable with the normal distribution $N(ab, b^2\sigma_a^2+a^2\sigma_b^2)$. Note that $b^2\sigma_a^2+a^2\sigma_b^2$ equals the approximated variance for $\hat{a}\hat{b}$ through the Delta method. The approximation needs only one of the variances goes to 0. Furthermore, $\hat{a}$ can be estimated from samples of (X, M) even when the outcome Y is not observed, as long as the samples are believed to be collected from the same population. Based on the limit distribution of $\hat{a} \times \hat{b}$, the sample size needed for a certain effect size and power can be calculated.

Next we consider the situation when the correlation between $\hat{a}$ and $\hat{b}$ are not 0. Assume the correlation coefficient is ρ. Therefore $(\hat{a}, \hat{b})$ has a bivariate normal distribution with the covariance $\rho\sigma_a\sigma_b$. Let $\delta(a) = \frac{a}{\sigma_a}$ and $\delta(b) = \frac{b}{\sigma_b}$ be the inverse of the coefficients of variation. It has been derived by [14] that the moment generating function is

$$M_Z(t) = \frac{exp[\frac{(\delta_a^2+\delta_b^2-2\rho\delta_a\delta_b)t^2+2\delta_a\delta_b t}{2(1-(1+\rho)t)(1-(1-\rho)t)}]}{((1-(1+\rho)t)(1-(1-\rho)t))^{1/2}}.$$

[14] has discussed that the Skewness of the distribution function is $(-2\sqrt{2}, 2\sqrt{2})$, and the maximum kurtosis value is 12. The density function is not likely to be normal. The power analysis becomes difficult.

Moreover, if a nonlinear method is used for the third-variable analysis, the power analysis is more complicated. We suggest to use simulation studies to calculate the sample size needed under specific settings.

12.2.2 Simulation Method

Based on linear models, the power of identifying an important third-variable depends on the following factors:

1. The coefficient, denoted as bx, of the exposure variable when it is used to estimate the third-variable;

2. The coefficient, denoted as bm, of the third-variable when it joins the exposure and other variables to estimate the outcome;

3. The variance of the exposure variable, denoted as sx;

4. The variance of the third-variable that cannot be explained by the exposure and other covariates, denoted as sm;

5. The variance of the outcome that cannot be explained by all variables, denoted as sy;

6. The R^2 of the regression function when the exposure is explained by other covariates;

7. The R^2 of the regression function when the third-variable is explained by other covariates;

8. The amount of the direct effect, denoted as c.

In the list, items 1 and 2 decide the size of the indirect effect. All other items decide the variance of the indirect effect. When there are only one third-variable, one exposure variable and one outcome, the following codes give an example of using simulations to find out the power of identifying the third-variable at certain sample size. In the example, all variables are continuous.

```
bx=    #1
bm=    #2
sx= #3
sm= #4
sy= #5
c= #8
nsim=100  #number of simulations for the power analysis
alpha=0.05 #
result1=NULL
set.seed(1)
for (i in 1:nSims){
  pred=rnorm(n,0,sx)
  e1=rnorm(n,0,sm)
  e2=rnorm(n,0,sy)
```

```
#continuous
m=a*pred+e1
y=c*x+b*m+e2

temp2<-mma(data.frame(m=m),y,pred=pred,mediator=1,
        jointm=list(n=1,j1=1),n2=100)
#m is forced in as a third-variable
D=summary(temp2,plot=F,alpha=alpha)
result1=rbind(result1,c(D$results$indirect.effect[[1]][,2]))}
```

To calculate the power, we check what is the proportion of times that the confidence interval does not include 0 out of the *nsim* simulations. If the power is too low, the sample size needs to be increased to try to reach the desired power.

We fixed the direct effect at zero and sx at 1, and did series of simulations to see how power is related with the specified parameters. Specifically, we allow the coefficients, bx and bm, to be chosen from the set $\{0.1, 0.3, 0.5, 1, 2\}$, the variances, sm or sy, from the set $\{0.5, 1, 2\}$, the significance level, *alpha*, to be $\{0.01, 0.05, 0.1\}$ and the sample sizes, n, ranges in $\{20, 50, 75, 100\}$. With the sets of parameters, we use the simulations to find out power at each combination of parameters. Finally, we fit a response-surface regression, where the calculated power is the outcome, all parameters, their quadratic form, and their two-way interactions are used as predictors. The results show that $bx, bm, sm, sy, alpha, n, bx^2, bm^2, bx \times bm, bx \times sm, bm \times sm$ are significant in predicting the power. Figure 12.1 shows the contour figure of every two parameters versus the fitted power. In the figure, when drawing figures of other two parameters, *alpha* is set at 0.05, n is at 50 and all other parameters are at 1. Figure 12.1 depicts how the power changes with the parameters.

Finally, if the outcome or third-variable is binary or time-to-event and/or other predictive models are needed for the third-variable analysis, the simulation can be adjusted to calculate powers.

12.3 Sequential Third-Variable Analysis and Third-Variable Analysis with Longitudinal Data

In general, third-variable effects can be presented in graphical models as sets of triangles that share one edge with two vertices. Vertices and edges from outside of the triangle set can be added to indicate potential covariate and moderate effects.

Sometimes, the third-variable effects cannot be simply demonstrated using a single set of triangle plots. If the graphical model can be decomposed into multiple sets of triangles with shared vertices, sequential third-variable analyses can be applied to each triangle sets. A sequential of third-variable

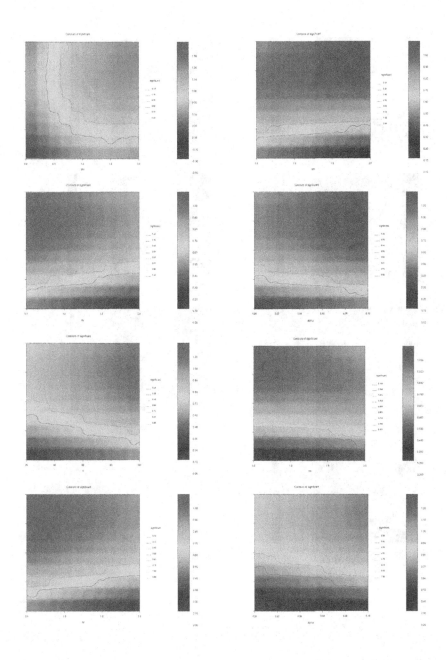

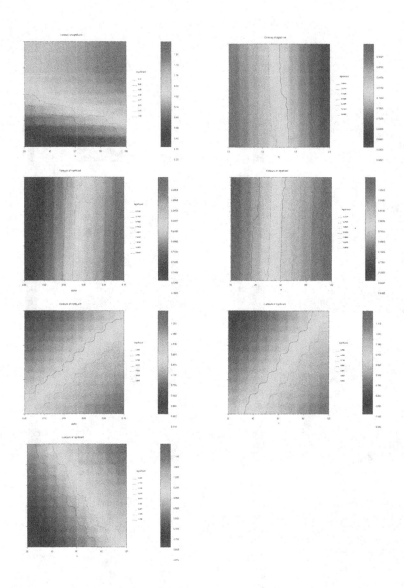

FIGURE 12.1
The contour plots of the power versus every two parameters.

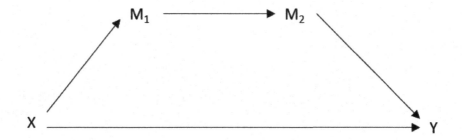

FIGURE 12.2
A third-variable, M1, is causal prior of the other third-variable, M2.

analysis can be performed to explore relationships among variables. Otherwise, it might be more convenient to use other modeling methods such as the structural equation models for the analysis.

An example of the situation is in Figure 12.2. In the figure, the third-variable M_1 is causally prior the other third-variable M_2. For such case, we can find the joint effect of M_1 and M_2 on the $X \to Y$ relationship. The relationship between M_1 and M_2 can be found by fitting a separate model.

Appendices

SAS Macros for Chapter 5

All these macros are provided in the book website.

setup_mma_macro.sas

```
/* ************************************* */
/* File Name: setup_mma_macro           */
/* Desciption:    */
/* Define Location for current analysis */
/* and define data names all parameters */
/* needed for mma package function      */
/* ************************************* */

options formchar="|----|+|---+=|-/\<>*";
proc options option=RLANG value; run;
*allows one to run R in SAS;

libname lib "Z:\Documents\LSU\LSU_RESEARCH\Yu";
%let path=Z:\Documents\LSU\LSU_RESEARCH\Yu;
%let pathd=Z:\Documents\LSU\LSU_RESEARCH\Yu;
*path is where the SAS dataset is stored;
%let data=weight_behavior;
*SAS dataset that is saved in the folder specified in pathd;
%let pre=lib.;
/*lib defines the libname*/
%let path_r=Z:/Documents/LSU/LSU_RESEARCH/Yu;

/*for data.org */
%let mediator=;
%let contmed=c(7:9,11:12);
%let binmed=c(6,10);
%let binref=c(1,1);
%let catmed=5;
%let catref=1;
%let predref=;
%let jointm=;
%let refy=;
```

```
%let alpha=0.4;
%let alpha2=0.4;
%let x=%str(data[,c(2,4:14)]);
%let pred=%str(data[,3]);
%let y=%str(data[,15]);

*if y is survival;
%let time=%str(data[]);
*days to last follow-up;
%let status=%str();
*0 or 1 indicator funciton;
%let y=%str(Surv(time,status));

%include "&path\Proc_R_dataorg.sas";
***data.org macro - save data.bin R data;
%Proc_R_dataorg(&pre,&path_r,&data,&mediator,&contmed,
&binmed,&binref,&catmed,&catref,&predref,&jointm,
&refy,&alpha,&alpha2,&x,&pred,&y,&time,&status);

/*for med*/
%let rdata=data.bin;
%let margin=;
%let D=;
%let distn=;
%let refy=;
%let n=2;
%let nu=;
%let nonlinear=FALSE;
%let df1=;
%let type=;

%include "&path\Proc_R_med.sas";

***med macro - print summary and save plot;
%Proc_R_med(&pre,&path_r,&rdata,&margin,&D,&distn,&refy,
&n,&nu,&nonlinear,&df1,&type)

/*for boot.med */
%let rdata=data.bin;
%let margin=;
%let D=;
%let distn=;
%let refy=;
%let n=2;
%let n2=4;
%let nu=;
%let nonlinear=TRUE;
%let df1=;
%let type=;
```

```
%let RE=;

%include "&path\Proc_R_bootmed.sas";

***boot.med macro - print summary and save plot;
%Proc_R_bootmed(&pre,&path_r,&rdata,&margin,&D,&distn,
&refy,&n,&n2,&nu,&nonlinear,&df1,&type,&RE);


/*for boot.med PLOT*/
%let vari=exercises;
%let alpha=;
%let quantile=;
%let xlim=c(1,70);

%include "&path\Proc_R_bootmed_plot.sas";

***boot.med PLOT macro - print summary and save plot;
%Proc_R_bootmed_plot(&pre,&path_r,&vari,&alpha,&quantile,&xlim);
```

R_submit_dataorg.sas

```
/* ************************************/
/* File Name: R_submit.sas            */
/* Desciption:     */
/* Run R in SAS     */
/* ************************************/
submit pre path_r data mediator contmed binmed binref catmed
catref predref jointm refy alpha alpha2 x pred y time status/R;

###load pacakges
library(mma)
library(survival)
data=read.table("&path_r/&data.txt",sep="\t", header = TRUE,
strip.white = FALSE)
attach(data)
x=&x
pred=&pred
status=&status
time=&time
y=&y

data.bin<-data.org(x=x,y=y,pred=pred,mediator=&mediator,
contmed=&contmed,binmed=&binmed,binref=&binref,
catmed=&catmed,catref=&catref,predref=&predref,
jointm=&jointm,refy=&refy,alpha=&alpha,alpha2=&alpha2)
save(data.bin, file = "&path_r/data.bin.RData")
out <- capture.output(data.bin)
cat("", out, file="&path_r/data_bin.txt", sep="n",
```

```
append=TRUE)
detach(data)

summary(data.bin)

endsubmit;
```

Proc_R_dataorg.sas

```
/* **********************************/
/* File Name: PROC_R_dataorg        */
/* Desciption:      */
/* Macro to save data.bin as txt file & */
/* R data   */
/* **********************************/

%macro proc_R_dataorg(pre,path_r,data,mediator,contmed,binmed,
binref,catmed,catref,predref,jointm,refy,alpha,alpha2,x,pred,
y,time,status);
/*write SAS dataset into a working directory*/
data wd;
set &pre.&data.;
run;

/*export SAS dataset to txt file*/
proc export data=wd
    outfile="&pathd\&data..txt"
        dbms=TAB replace;
    putnames=YES;
run;

proc iml;
pre="&pre";
path="&path";
path_r="&path_r";
data="&data";
mediator="&mediator";
contmed="&contmed";
binmed="&binmed";
binref="&binref";
catmed="&catmed";
catref="&catref";
predref="&predref";
jointm="&jointm";
refy="&refy";
alpha="&alpha";
alpha2="&alpha2";
x="&x";
pred="&pred";
```

```
y="&y";
time="&time";
status="&status";
%INCLUDE  "&path\R_submit_dataorg.sas";
%mend proc_R_dataorg;
```

Proc_R_med.sas

```
/* **********************************/
/* File Name: PROC_R_med              */
/* Desciption:    */
/* Macro to summarize mediation effects */
/*       */
/* R data   */
/* **********************************/

%macro Proc_R_med(pre,path_r,rdata,margin,D,distn,refy,n,
nu,nonlinear,df,type);

proc iml;
pre="&pre";
path_r="&path_r";
rdata="&rdata";
margin="&margin";
D="&D";
distn="&distn";
refy="&refy";
n="&n";
nu="&nu";
nonlinear="&nonlinear";
df="&df";
type="&type";
%INCLUDE  "&path_r\R_submit_med.sas";

%mend Proc_R_med;
```

R_submit_med.sas

```
/* **********************************/
/* File Name: R_submit_med.sas        */
/* Desciption:     */
/* Run R in SAS - med */
/* **********************************/
submit pre path_r rdata margin D distn refy n nu nonlinear df
 type/R;
###load pacakges
library(mma)
load("&path_r/&rdata.RData")
```

```
temp.med<-med(data=data.bin,margin=&margin,D=&D,distn=&distn,
refy=&refy,n=&n,nu=&nu,nonlinear=&nonlinear,df=&df,type="&type")
save(temp.med, file = "&path_r/temp.med.RData")
print(temp.med)
temp.med[["model"]]
endsubmit;
```

Proc_R_bootmed.sas

```
/* **********************************/
/* File Name: PROC_R_bootmed         */
/* Desciption:    */
/* Macro to summarize mediation effects */
/* and save plot data_bin_plot.png       */
/* R data  */
/* **********************************/


%macro Proc_R_bootmed(pre,path_r,rdata,margin,D,distn,
refy,n,n2,nu,nonlinear,df1,type,RE);

proc iml;
pre="&pre";
path_r="&path_r";
rdata="&rdata";
margin="&margin";
D="&D";
distn="&distn";
refy="&refy";
n="&n";
n2="&n2";
nu="&nu";
nonlinear="&nonlinear";
df1="&df1";
type="&type";
RE="&RE";
%INCLUDE  "&path_r\R_submit_bootmed.sas";

%mend Proc_R_bootmed;
```

R_submit_bootmed.sas

```
/* **********************************/
/* File Name: R_submit_bootmed.sas      */
/* Desciption:   */
/* Run R in SAS - bootmed */
/* **********************************/
submit pre path_r rdata margin D distn refy n n2 nu
```

```
nonlinear df1 type RE/R;
###load pacakges
library(mma)
load("&path_r/&rdata.RData")
data.bin.plot<-boot.med(data=data.bin,margin=&margin,D=&D,
distn=&distn,refy=&refy,n=&n,n2=&n2,nu=&nu,
nonlinear=&nonlinear,df1=&df1,type="&type")
png("&path_r/data_bin_plot.png")
summary(data.bin.plot, RE=&RE)
dev.off()
save(data.bin.plot, file = "&path_r/data.bin.plot.RData")

endsubmit;
```

Proc_R_plot.sas

```
/* **********************************/
/* File Name: PROC_R_Plot          */
/* Desciption:  */
/* Macro to save plot data_bin_plot2.png */
/* R data  */
/* **********************************/


%macro Proc_R_plot(pre,path_r,vari,alpha,quantile,xlim) ;

proc iml;
pre="&pre";
path_r="&path_r";
vari="&vari";
alpha="&alpha";
quantile="&quantile";
xlim="&xlim";
%INCLUDE  "&path_r\R_submit_plot.sas";

%mend Proc_R_plot;
```

R_submit_plot.sas

```
/* **********************************/
/* File Name: R_submit_plot.sas    */
/* Desciption: */
/* Run R in SAS - plot */
/* **********************************/
submit pre path_r vari alpha quantile xlim/R ;
###load pacakges
library(mma)
load("&path_r/data.bin.plot.RData")
```

```
png("&path_r/data_bin_plot2.png")
plot(data.bin.plot, vari="&vari",xlim=&xlim,alpha=&alpha,
quantile=&quantile)
dev.off()

endsubmit;
```

R-codes for simulations in Chapter 6

Codes to generate data for Assumption 1

```
library(mma)
library(MASS)
a0=0.2
a1=0.1

a=0.773
b=0.701
n=50
nSims=100
result1=NULL
result2=NULL

set.seed(13)
for (rho in c(-0.9, -0.5, -0.1, 0, 0.1, 0.5, 0.9))
 for (c1 in c(0, 0.518, 1.402, 2.150))
  for (cy in c(0, 0.259, 0.701, 1.705))
    for (i in 1:nSims){
      e1=rnorm(n,0,1)
      e2=rnorm(n,0,1)
      sigma=matrix(c(1,rho,rho,1),2,2)
      G=mvrnorm(n=50,rep(0,2),sigma)
      X1=G[,1]
      C=G[,2]
      pred=X1

      X=a0+a*pred+e1
      Y=a1+b*X+c1*pred+cy*C+e2

      Xo=data.frame(X)
      X2=data.frame(X,C)
      Yo=data.frame(Y)
      pred1=data.frame(X1)

      #linear results
```

```
     temp2<-mma(Xo,Yo,pred=pred1,mediator=1,jointm=list(n=1,
        j1=1),alpha=0.1,alpha2=0.1)
#M is forced in as a third-variable
     D=summary(temp2,plot=F)
     result1=rbind(result1,c(a,b,c1,rho,cy,i,D$results$total.effect,
   D$results$direct.effect,D$results$indirect.effect[[1]][,2]))

     #nonlinear results
     temp2<-mma(Xo,Yo,pred=pred1,mediator=1,jointm=list(n=1,j1=1),
        alpha=0.1,alpha2=0.1,nonlinear=T)
     D=summary(temp2,plot=F)
     result2=rbind(result2,c(a,b,c1,rho,cy,i,D$results$total.effect,
   D$results$direct.effect,D$results$indirect.effect[[1]][,2]))
}

# The right model if C is treated as a covariate
     temp2<-mma(X2,Yo,pred=pred1,mediator=1,jointm=list(n=1,
        j1=1),alpha=0.1,alpha2=0.1)

# The right model if C is treated as another third variable
     temp2<-mma(X2,Yo,pred=pred1,mediator=1:2,jointm=list(n=1,
        j1=1),alpha=0.1,alpha2=0.1)
```

Codes to generate data for Assumption 2

```
b=0.701
c1=1.402
n=50
nSims=100
result1=NULL
result2=NULL

set.seed(14)
for (rho in c(-0.9, -0.5, -0.1, 0, 0.1, 0.5, 0.9))
  for (a in c(0,0.286,0.773,1.185))
    for (cm in c(0, 0.259, 0.701, 1.705))
      for (i in 1:nSims){
        e1=rnorm(n,0,1)
        e2=rnorm(n,0,1)
        sigma=matrix(c(1,rho,rho,1),2,2)
        G=mvrnorm(n=50,rep(0,2),sigma)
        X1=G[,1]
        C=G[,2]
        pred=X1

        X=a0+a*pred+cm*C+e1
        Y=a1+b*X+c1*pred+e2
```

```
        Xo=data.frame(X)

        Yo=data.frame(Y)
        pred1=data.frame(pred)
        Co=data.frame(C)
        #linear results
        temp2<-mma(Xo,Yo,pred=pred1,mediator=1,jointm=list(n=1,j1=1),
alpha=0.1,alpha2=0.1)
        D=summary(temp2,plot=F)
        result1=rbind(result1,c(a,b,c1,rho,cm,i,D$results$total.effect,
             D$results$direct.effect,
                    D$results$indirect.effect[[1]][,2]))

        #nonlinear results
        temp2<-mma(Xo,Yo,pred=pred1,mediator=1,jointm=list(n=1,j1=1),
            alpha=0.1,alpha2=0.1,nonlinear=T)
        D=summary(temp2,plot=F)
        result2=rbind(result2,c(a,b,c1,rho,cm,i,D$results$total.effect,
             D$results$direct.effect,
                    D$results$indirect.effect[[1]][,2]))
    }

# The right model
        temp2<-mma(Xo,Yo,pred=pred1,mediator=1, cova=data.frame(C),
            jointm=list(n=1,j1=1),alpha=0.1,alpha2=0.1)
```

Codes to generate data for Assumption 3

```
a0=0.2
a1=0.1

a=0.773
c1=0.701
n=50
nSims=100
result1=NULL
result2=NULL

set.seed(2)
for (rho in c(-0.9, -0.5, -0.1, 0, 0.1, 0.5, 0.9))
  for (b in c(0,0.518,1.402,2.150))
    for (cy in c(0,0.259,0.701,1.705))
      for (i in 1:nSims){
        e1=rnorm(n,0,1)
        e2=rnorm(n,0,1)
        sigma=matrix(c(1,rho,rho,1),2,2)
        G=mvrnorm(n=50,rep(0,2),sigma)
        e3=G[,1]
        C=G[,2]
```

```
        pred=e1

        X=a0+a*pred+e3
        Y=a1+b*X+c1*pred+cy*C+e2

        Xo=data.frame(X)
        X2=data.frame(X,C)
        Yo=data.frame(Y)
        pred1=data.frame(pred)

        #linear results
        temp2<-mma(Xo,Yo,pred=pred1,mediator=1,jointm=list(n=1,j1=1),
            alpha=0.1,alpha2=0.1)
        D=summary(temp2,plot=F)
        result1=rbind(result1,c(a,b,c1,rho,cy,i,D$results$total.effect,
                D$results$direct.effect,
D$results$indirect.effect[[1]][,2]))

        #nonlinear results
        temp2<-mma(Xo,Yo,pred=pred1,mediator=1,jointm=list(n=1,j1=1),
            alpha=0.1,alpha2=0.1,nonlinear=T)
        D=summary(temp2,plot=F)
        result2=rbind(result2,c(a,b,c1,rho,cy,i,D$results$total.effect,
                D$results$direct.effect,
                D$results$indirect.effect[[1]][,2]))
    }
#codes to fit the right model
        temp2<-mma(X2,Yo,pred=pred1,mediator=1,jointm=list(n=1,j1=1),
            alpha=0.1,alpha2=0.1)
```

Codes to generate data for Assumption 4

```
a1=0.773
b1=0.701
c1=1.250
n=50
nSims=100
result1=NULL

set.seed(5)
for (cm in c(-0.9, -0.5, -0.1, 0, 0.1, 0.5,0.9))
  for (a2 in c(0,0.286,0.773,1.185))
    for (b2 in c(0,0.259,0.701,1.705))
      for (i in 1:nSims){
        pred=rnorm(n,0,1)
        e2=rnorm(n,0,1)
        e3=rnorm(n,0,1)
        e4=rnorm(n,0,1)
```

```
M2=a2*pred+e2

M1=a1*pred+cm*M2+e3

Y=b1*M1+c1*pred+b2*M2+e4

Xo=data.frame(M1,M2)
Yo=data.frame(Y)
pred1=data.frame(pred)

#linear results
temp1<-mma(Xo,Yo,pred=pred1,mediator=1:2,jointm=list(n=1,j1=1:2),
    alpha=0.1,alpha2=0.1)
D1=summary(temp1,plot=F)
result1=rbind(result1,c(a,b,c1,d,cm,cy,i,D1$results$total.effect,
        D1$results$direct.effect,
                D1$results$indirect.effect[[1]][,2],
        D1$results$indirect.effect[[1]][,3],
        D1$results$indirect.effect[[1]][,4]))
}
```

Bibliography

[1] JM Albert. Mediation analysisvia potential outcomes models. *Statistics in Medicine*, 27(8):1282–1304, 2008.

[2] HR Alker. *A Typology of Ecological Fallacies, In Quantitative Ecological Analysis in the Social Sciences*. The MIT Press, Cambridge, MA, 1969.

[3] Duane F Alwin and Robert M Hauser. The decomposition of effects in path analysis. *American Sociological Review*, 40(1):37, Feb 1975.

[4] Reuben M Baron and David A Kenny. The moderator–mediator variable distinction in social psychological research: Conceptual, strategic, and statistical considerations. *Journal of Personality and Social Psychology*, 51(6):1173–1182, 1986.

[5] DJ Bauer. Estimating multilevel linear models as structureal equatin models. *Journal of Educational and Behavioral Statistics*, 28(2):135–167, 2003.

[6] Daniel J Bauer, Kristopher J Preacher, and Karen M Gil. Conceptualizing and testing random indirect effects and moderated mediation in multilevel models: New procedures and recommendations. *Psychological Methods*, 11(2):142–163, 2006.

[7] BB Bevers, Mark Helvie, Ermelinda Bonaccio, Kristine E Calhoun, Mary B Daly, William B Farrar, Judy E Garber, Richard Gray, Caprice C Greenberg, Rachel Greenup, Nora M Hansen, Randall E Harris, Alexandra S Heerdt, Teresa Helsten, Linda Hodgkiss, Tamarya L Hoyt, John G Huff, Lisa Jacobs, Constance Dobbins Lehman, Barbara Monsees, Bethany L Niell, Catherine C Parker, Mark Pearlman, Liane Philpotts, Laura B Shepardson, Mary Lou Smith, Matthew Stein, Lusine Tumyan, Cheryl Williams, Mary Anne Bergman, and Rashmi Kumar. Breast cancer screening and diagnosis clinical practice guidelines in oncology. *Journal of the National Comprehensive Cancer Network*, 1(2):242–242, Apr 2003.

[8] L Breiman. Regression trees. In *Classification and Regression Trees*, pages 216–265. Routledge, 1984.

[9] L Breiman. Prediction Games and Arcing Algorithms. *Neural Computation*, 11(7):1493–1517, Oct 1999.

[10] L Breiman, J Friedman, R Olshen, and C Stone. *Classification and Regression Trees*. Wadsworth, New York, 1984.

[11] Wolfgang Brunauer, Stefan Lang, and Nikolaus Umlauf. Modelling house prices using multilevel structured additive regression. *Statistical Modelling*, 13(2):95–123, 2013.

[12] F Busing. Distribution characteristics of variance estimates in two-level models. *Unpublish manuscript*, 1993.

[13] Wentao Cao, Yaling Li, and Qingzhao Yu. Sensitivity analysis for assumptions of general mediation analysis. *Communications in Statistics - Simulation and Computation*, 2021.

[14] Cecil C Craig. On the frequency function of xy. *The Annals of Mathematical Statistics*, 7(1):1–15, Mar 1936.

[15] JR Edwards and LS Lambert. Methods for integrating moderation and mediation: a general analytical framework using moderated path analysis. *Psychological methods*, 12(1):1–22, Mar 2007.

[16] Bradley Efron and RJ Tibshirani. *An Introduction to the Bootstrap*. Chapman and Hall/CRC, May 1994.

[17] EH Arthur and WK Robert. Ridge regression: Applications to nonorthogonal problems. *Technometrics*, 12(1):55–67, Feb 1970.

[18] AJ Fairchild and DP MacKinnon. A general model for testing mediation and moderation effects. *Prev Sci*, 10(12):87–99, 2009.

[19] Y Fan. Multiple mediation analysis with general predictive models. *Ph.D. dissertation*, 2012.

[20] Paige Fisher, Wentao Cao, and Qingzhao Yu. Using SAS macros for multiple mediation analysis in r. *Journal of Open Research Software*, 8(1):22, Oct 2020.

[21] Y Freund and R Schapire. Experiments with a new boosting algorithm. In *Proceedings of the Thirteenth International Conference on Machine Learning*, pages 148–156. Morgan Kaufmann, July 1996.

[22] J Friedman. Greedy function approximation: a gradient boosting machine. *The Annals of Statistics*, 29(5):1189–1536, Oct 2001.

[23] J Friedman. Stochastic gradient boosting. *Computational Statistics & Data Analysis*, 38(4):367–378, Feb 2002.

[24] J Friedman, T Hastie, and R Tibshirani. Additive logistic regression: a statistical view of boosting. *The Annals of Statistics*, 28(2):337–407, Apr 2000.

[25] JH Friedman and W Stuetzle. Estimating optimal transformations for multiple regression and correlation. *Journal of the American Statistical,* 76:817–823, 1981.

[26] JH Friedman and JJ Meulman. Multiple additive regression trees with application in epidemiology. *Statistics in Medicine,* 22(9):1365–1381, 2003.

[27] JH Friedman and BE Popescu. Predictive learning via rule ensembles. *The Annals of Applied Statistics,* 2(3):916–954, Sep 2008.

[28] Andrew Gelman. Multilevel (hierarchical) modeling: What it can and cannot do. *Technometrics,* 48(3):432–435, 2006.

[29] Alix I. Gitelman. Estimating causal effects from multilevel group-allocation data. *Journal of Educational and Behavioral Statistics,* 30(4):397–412, 2005.

[30] Michael J Greenwald and Marlon G Boarnet. Built environment as determinant of walking behavior: Analyzing nonwork pedestrian travel in portland, oregon. *Transportation Research Record: Journal of the Transportation Research Board,* 1780(1):33–41, Jan 2001.

[31] T Hastie, R Tibshirani, and J Friedman. *The Elements of Statistical Learning.* Springer, 2009.

[32] TJ Hastie and RJ Tibshirani. *Generalized Additive Models.* Chapman & Hall/CRC, 1990.

[33] A Hayes and NJ Rockwood. Regression-based statistical mediation and moderation analysis in clinical research: Observations, recommendations, and implementation. *Behaviour Research and Therapy,* 98:39–57, Nov 2017.

[34] AF Hayes. *Introduction to mediation, moderation, and conditional process analysis: A regression-based approach (2nd Ed.).* New York: The Guilford Press, 2018.

[35] Doug Hemken. Using r from sas, 2017.

[36] PW Holland, C Glymour, and C Granger. Statistics and causal inference. *ETS Research Report Series,* 1985(2):i–72, Dec 1986.

[37] R.t. Hurt and et al. The obesity epidemic: Challenges, health initiatives, and implications for gastroenterologists. *Gastroenterology & Hepatology,* 6:780–792, 2010.

[38] HH Hyman. *Survey Design and Analysis: Principles, Cases and Procedures.* Free Press of Glencoe, Glencoe, 1955.

[39] K Imai and T Yamamoto. Identification and sensitivity analysis for multiple causal mechanisms: Revisiting evidence from framing experiments. *Political Analysis*, 21(02):141–171, 2013.

[40] J Jasem, A Amini, R Rabinovitch, VF Borges, A Elias, CM Fisher, and P Kabos. 21-gene recurrence score assay as a predictor of adjuvant chemotherapy administration for early-stage breast cancer: An analysis of use, therapeutic implications, and disparity profile. *Journal of Clinical Oncology*, 34(17):1995–2002, Jun 2016.

[41] Charles M Judd and David A Kenny. Process analysis. *Evaluation Review*, 5(5):602–619, Oct 1981.

[42] David A Kenny, Niall Bolger, and Josephine D Korchmaros. Lower level mediation in multilevel models. *Psychological Methods*, 8(2):115–128, 2003.

[43] Jae-On Kim and G Donald Ferree. Standardization in causal analysis. *Sociological Methods & Research*, 10(2):187–210, Nov 1981.

[44] GS Kimeldorf and G Wahba. A correspondence between bayesian estimation on stochastic processes and smoothing by splines. *The Annals of Mathematical Statistics*, 41(2):495–502, 1970.

[45] JL Krull and DP MacKinnon. Multilevel modeling of individual and group level mediated effects. *Multivariate Behavioral Research*, 36(2):249–277, 2001.

[46] Stefan Lang, Nikolaus Umlauf, Peter Wechselberger, Kenneth Harttgen, and Thomas Kneib. Multilevel structured additive regression. *Statistics and Computing*, 24(2):223–238, 2012.

[47] Bin Li, Qingzhao Yu, Lu Zhang, and Meichin Hsieh. Regularized multiple mediation analysis. *Statistics and Its Interface*, 14(4):449–458, 2021.

[48] David Lunn. *The BUGS Book*. Chapman and Hall/CRC, Boca Raton, FL, Oct 2012.

[49] David P MacKinnon, Ghulam Warsi, and James H Dwyer. A simulation study of mediated effect measures. *Multivariate Behavioral Research*, 30(1):41–62, Jan 1995.

[50] Melissa C Nelson, Penny Gordon-Larsen, Yan Song, and Barry M Popkin. Built and social environments. *American Journal of Preventive Medicine*, 31(2):109–117, Aug 2006.

[51] US Department of Health and Human Services. National diabetes statistics report: estimates of diabetes and its burden in the united states. 2014. *Atlanta, GA: Centers for Disease Control and Prevention*, 2014.

[52] R Otero-Sabogal and et al. Dietary practices, alcohol consumption, and smoking behavior: ethnic, sex, and acculturation differences. *J Natl Cancer Inst Monogr*, 18:73–82, 1995.

[53] P Gordon-Larsen, P Page MC Nelson, and BM Popkin. Inequality in the built environment underlies key health disparities in physical activity and obesity. *Pediatrics*, 117:417–424, 2006.

[54] S Paik, S Shak, G Tang, and et al. A multigene assay to predict recurrence of tamoxifen-treated, node-negative breast cancer. *New England Journal of Medicine*, 351(27):2817–2826, Dec 2004.

[55] S Paik, G Tang, S Shak, and et al. Gene expression and benefit of chemotherapy in women with node-negative, estrogen receptor–positive breast cancer. *Journal of Clinical Oncology*, 24(23):3726–3734, Aug 2006.

[56] J Palmgren. Regression models for bivariate binary response. *University of Washington Biostatistics Working Paper Series*, 1989.

[57] J Pearl. Direct and indirect effects. *Proceedings of the Seventeenth Conference on Uncertainty and Artificial Intelligence*, pages 411–420, Oct 2001.

[58] R Perez-Escamilla and P Putnik. The role of acculturation in nutrition, lifestyle, and incidence of type 2 diabetes among latinos. *J Nutr*, 137:860–870, 2007.

[59] VI Petkov, DP Miller, N Howlader, and et al. Breast-cancer-specific mortality in patients treated based on the 21-gene assay: a SEER population-based study. *npj Breast Cancer*, 2(1), Jun 2016.

[60] Keenan A Pituch, Tiffany A Whittaker, and Laura M Stapleton. A comparison of methods to test for mediation in multisite experiments. *Multivariate Behavioral Research*, 40(1):1–23, 2005.

[61] DJ Press, A Ibraheem, ME Dolan, KH Goss, S Conzen, and D Huo. Racial disparities in omission of oncotype DX but no racial disparities in chemotherapy receipt following completed oncotype DX test results. *Breast Cancer Research and Treatment*, 168(1):207–220, Nov 2017.

[62] SW Raudenbush and AS Bryk. *Hierarchical Linear Models: Applications and Data Analysis Methods (2nd edition)*. Sage, Thousand Oaks, CA, 2002.

[63] LJ Ricks-Santi and JT McDonald. Low utility of oncotype DX® in the clinic. *Cancer Medicine*, 6(3):501–507, Feb 2017.

[64] JM Robins and S Greenland. Identifiability and exchangeability for direct and indirect effects. *Epidemiology*, 3(2):143–155, Mar 1992.

[65] Donald B Rubin. Estimating causal effects of treatments in randomized and nonrandomized studies. *Journal of Educational Psychology*, 66(5):688–701, 1974.

[66] A Rundle, KM Neckerman, L Freeman, GS Lovasi, M Purciel, J Quinn, C Richards, N Sircar, and C Weiss. Neighborhood food environment and walkability predict obesity in new york city. *Environmental Health Perspectives*, 117:442, 2009.

[67] Brian E Saelens, James F Sallis, and Lawrence D Frank. Environmental correlates of walking and cycling: Findings from the transportation, urban design, and planning literatures. *Annals of Behavioral Medicine*, 25(2):80–91, Apr 2003.

[68] RJ Sampson and JD Morenoff. Spatial (dis)advantage and homicide in chicago neighborhoods. *Spatially Integrated Social Science*, 2004.

[69] R Schapire. The strength of weak learnability. *Machine Learning*, 5(2):197–227, July 1990.

[70] N Simon, J Friedman, T Hastie, and R Tibshirani. Regularization paths for cox's proportional hazards model via coordinate descent. *Journal of Statistical Software*, 39(5), 2011.

[71] Tom AB Snijders and Roel J Bosker. *Multilevel Analysis: An Introduction to Basic and Advanced Multilevel Modeling*. Sage, Thousand Oaks, CA, 1999.

[72] TR Ten Have, MM Joffe, KG Lynch, GK Brown, SA Maisto, and AT Beck. Causal mediation analyses with rank preserving models. *Biometrics*, 63(3):926–934, Mar 2007.

[73] R Tibshirani. Regression shrinkage and selection via the lasso. *Journal of the Royal Statistical Society: Series B (Statistical Methodology)*, 58(1):267–288, Jan 1996.

[74] Davood Tofighi, Stephen G West, and David P MacKinnon. Multilevel mediation analysis: the effects of omitted variables in the 1-1-1 model. *British Journal of Mathematical and Statistical Psychology*, 66(2):290–307, 2012.

[75] TJ Vanderweele. *Explanation in Causal Inference – Methods for Mediation and Interaction*. Oxford, New York, 2015.

[76] TJ VanderWeele. Marginal structural models for the estimation of direct and indirect effects. *Epidemiology*, 20(1):18–26, Jan 2009.

[77] TJ VanderWeele and WT Robinson. On the causal interpretation of race in regressions adjusting for confounding and mediating variables. *Epidemiology*, 25(4):473–484, Jul 2014.

[78] TJ Vanderweele and S Vansteelandt. Conceptual issues concerning mediation, interventions and composition. *Statistics and Its Interface*, 2(4):457–468, 2009.

[79] Q Yu, Y Fan, and X Wu. General multiple mediation analysis with an application to explore racial disparities in breast cancer survival. *Journal of Biometrics & Biostatistics*, 05(02), 2013.

[80] Q Yu and B Li. mma: An r package for multiple mediation analysis. *Journal of Open Research Software*, 5:11, 2017.

[81] Q Yu and B Li. *mmabig: Multiple Mediation Analysis for Big Data Sets.* CRAN, 2018.

[82] Q Yu and B Li. A multivariate multiple third-variable effect analysis with an application to explore racial and ethnic disparities in obesity. *Journal of Applied Statistics*, 48(4):750–764, 2021.

[83] Q Yu, B Li, and RA Scribner. Hierarchical additive modeling of nonlinear association with spatial correlations-an application to relate alcohol outlet density and neighborhood assault rates. *Statistics in Medicine*, 28(14):1896–1912, Jun 2009.

[84] Q Yu, KL Medeiros, X Wu, and RE Jensen. Nonlinear predictive models for multiple mediation analysis: With an application to explore ethnic disparities in anxiety and depression among cancer survivors. *Psychometrika*, 83(4):991–1006, Apr 2018.

[85] Q Yu, RA Scribner, C Leonardi, L Zhang, C Park, L Chen, and NR Simonsen. Exploring racial disparity in obesity: a mediation analysis considering geo-coded environmental factors. *Spatial and Spatio-temporal Epidemiology*, 21:13–23, 2017.

[86] Q Yu, X Wu, B Li, and RA Scribner. Multiple mediation analysis with survival outcomes: With an application to explore racial disparity in breast cancer survival. *Statistics in Medicine*, 38(3):398–412, Sep 2018.

[87] Qingzhao Yu and Bin Li. Third-variable effect analysis with multilevel additive models. *PLOS ONE*, 15(10):e0241072, Oct 2020.

[88] Qingzhao Yu, Lu Zhang, Xiaocheng Wu, and Bin Li. Inference on moderation effect with third-variable effect analysis – application to explore the trend of racial disparity in oncotype dx test for breast cancer treatment. *Journal of Applied Statistics*, 2021.

[89] Ying Yuan and David P MacKinnon. Bayesian mediation analysis. *Psychological Methods*, 14(4):301–322, 2009.

[90] L Zhang, M Hsieh, V Petkov, Q Yu, Y Chiu, and X Wu. Trend and survival benefit of oncotype dx use among female hormone receptor positive breast cancer patients in 14 seer registries, 2004-2015. *In preparation*, 2019.

[91] L Zhang, M Hsieh, V Petkov, Q Yu, Y Chiu, and X Wu. Trend and survival benefit of oncotype dx use among female hormone receptor positive breast cancer patients in 14 seer registries, 2004-2015. *Breast Cancer Research and Treatment*, 2020.

[92] Zhen Zhang, Michael J Zyphur, and Kristopher J Preacher. Testing multilevel mediation using hierarchical linear modesl, problems and solutions. *Organizational Research Methods*, 12(4):695–719, 2009.

[93] Yu, Q, Yu, M, Zou, J, Wu, X, Gomez, SL, Li, B. 2021. Multilevel Mediation Analysis on Time-to-Event Outcomes - Exploring racial/ethnic Disparities in Breast Cancer Survival in California. *Research Methods in Medicine & Health Sciences*. DOI: 10.1177/26320843211061292.

[94] H Zou and T Hastie. Regularization and variable selection via the elastic net. *Journal of the Royal Statistical Society: Series B (Statistical Methodology)*, 67(2):301–320, Apr 2005.

Index